William Scrope

Days of deer-stalking in the Scottish Highlands

including an account of the nature and habits of the red deer, a description of the Scottish

forests, and historical notes on the earlier field-sports of Scotland

William Scrope

Days of deer-stalking in the Scottish Highlands
including an account of the nature and habits of the red deer, a description of the Scottish forests, and historical notes on the earlier field-sports of Scotland

ISBN/EAN: 9783743345942

Manufactured in Europe, USA, Canada, Australia, Japa

Cover: Foto ©berggeist007 / pixelio.de

Manufactured and distributed by brebook publishing software (www.brebook.com)

William Scrope

Days of deer-stalking in the Scottish Highlands

THE FOREST JOUST.

Edwin Landseer R.A.

DAYS OF DEER-STALKING,

by

William Scrope, Esqre

LONDON HAMILTON ADAMS & C°
GLASGOW THOMAS L MORISON.
1883

DAYS OF DEER-STALKING

IN THE SCOTTISH HIGHLANDS

*Including an Account of the Nature and Habits of the Red Deer
a Description of the Scottish Forests, and Historical
Notes on the Earlier Field-Sports of Scotland.*

*With Highland Legends, Superstitions, Traditions, Folk-Lore, and
Tales of Poachers and Freebooters.*

BY

WILLIAM SCROPE, ESQ.

Author of "Days and Nights of Salmon Fishing"

ILLUSTRATED BY

SIR EDWIN AND CHARLES LANDSEER

LONDON: HAMILTON, ADAMS & CO.
GLASGOW: THOMAS D. MORISON
1883

Among the works published in connection with field-sports in Scotland, probably none have been more sought after than those two most interesting books by Mr. Scrope, namely, "Days of Deer-Stalking" and "Days and Nights of Salmon Fishing," and yet it may be said that no works in that departments of literature have been more difficult to get. Indeed, their scarcity has been so great as frequently, when inquired for, to cause the remark, that it is almost in vain to go in quest of either. Again, when copies do happen to come into the market, they command such prices as to startle persons of moderate ideas, notwithstanding the very high interest and intrinsic value attaching to them. Such considerations have suggested the republication of the first named book. Since its original issue some changes have taken place with regard to minor matters, but the great scope of the work is as interesting and fascinating as ever. The points of interest are exceedingly varied in character, and meet the tastes of a wide circle. Whether one's bent leads in the direction of field-sports, natural history, topography, antiquities, or historical lore in connection with the Highlands in general, the reader, more especially if he be interested in Scottish incident, will find a charm and a source of interesting information in the volume not surpassed in such respects by any work of the kind.

PREFACE.

"SHALL a poaching, hunting, hawking 'squire presume to trespass on the fields of literature?" These words, or others of similar import, I remember to have encountered in one of our most distinguished reviews. They ring still in my ears, and fill me with apprehension as it is; but they would alarm me much more if I had attempted to put my foot within the sacred enclosures alluded to. These are too full of spring traps for my ambition, and I see "*this is to give notice*" written in very legible characters, and take warning accordingly.

Literature?—Heaven help us!—far from it; I have no such presumption; I have merely attempted to describe a very interesting pursuit as nearly as possible in the style and spirit in which I have always seen it carried on. Ten years' successful practice in the forest of Atholl have enabled me to enter into all the details that are connected with deer-stalking. That it is a chase which throws all our other field-sports far in the back-ground, and, indeed makes them appear wholly insignificant, no one who has been initiated in it will attempt to deny. The beautiful motions of the deer, his picturesque and noble appearance, his sagacity, and the skilful generalship which can alone ensure success in the pursuit of him, keep the mind in a constant state of pleasurable excitement.

Those arts which are the most successful for killing the stag will apply to almost all other wild quadrupeds of the nobler sort; and a correct acquaintance with them might possibly be the means of saving many an adventurer's life, whose actual sustenance, and that of his companions, depended on his skill in hunting. In exploring unknown regions for the advancement of science, or cast, as men may be, on a desolate shore, how necessary, how indispensable, is a knowledge of the huntsman's craft for the actual preservation of existence! And yet, in such travels as I have read, I have never seen this craft fully explained, the adventurer having been under the guidance of the natives, and for the most part a novice in the business.

In my narrative of deer-stalking I have not, except in one instance, noted my best success—far from it. My aim has been to confine myself to such events as I thought best calculated to illustrate a diversion, which all sportsmen, who have the means in their power, are now pursuing with unabating ardour.

PREFACE.

I have thought it desirable to describe the motions of the red deer under every variety of pursuit and danger to them; to set forth their great sagacity and self-possession; their courage and noble bearing; the bay; the method in which they are prepared for being taken home; and many particulars relating to their natural history and habits.

I have attempted also to illustrate all the essential points that occur in stalking deer, both in slow and quick time, and to describe all the various turns and accidents of the chase drawn from actual experience. This, I thought, could be best done by the recital of moderate sport, since a long catalogue of deer, killed in succession on the same day, unaccompanied by some striking or unusual incident, would only be a tedious repetition of events similar to each other. In practice, however, I did my best, as fine venison was always in request. If my success was occasionally very considerable, it must be recollected that the deer were numerous, and that I was assisted by clever scouts. The being my own stalker, also, was an advantage that long practice enabled me to profit from: no one, I think, can make the best of events when his movements are controlled by others, and are a mystery to himself.

To the courtesy of the noblemen and gentlemen, proprietors of the various magnificent deer forests in Scotland, I am indebted for the short descriptions I have given of them, and they are inserted nearly as I received them, with a due and lasting sense of the honour and obligation that has been conferred upon me; the account of the forest of Atholl alone has been put together wholly by myself; with that I am pretty conversant, but not with the others. Whilst I am on this subject, I cannot avoid expressing a regret that the communications sent to me have not done sufficient justice to the scenery they treat of, which in its wild effect, and peculiar determination of character, is admirably suited to the disposition and pursuits of its brave and romantic inhabitants.

It will be seen how much I am indebted to Mr. Macneill, of Colonsay, for his very interesting account of the original Scotch greyhound, and for his picturesque description of the novel amusement of *deer-coursing*. I am myself unacquainted with the distinguishing characteristics of the ancient Scotch and Irish greyhound; but there are still many magnificent dogs in the possession of Scotch gentlemen and chiefs, however they may be descended; and a late celebrated sale will prove how highly some of the present breed are esteemed by the public.

I have to boast of two poetical contributions, from the

Hon. Henry T. Liddell, which appear to me to be exquisitely beautiful. Mr. D'Israeli likewise has ornamented my pages with some beautiful lines, paraphrased from a translation from the Gaelic, most obligingly sent to me by the Marquis of Breadalbane.

To my accomplished friend Mr. Skene, of Rubislaw, I am under very great obligations, not only for some valuable communications from himself, but also for other intelligence which I have obtained by his means, and through his influence.

The Duchess Countess of Sutherland has condescendingly procured for me a full account of her magnificent possessions in the North, which has been most ably put together by Mr. Taylor, to whose skill and diligence I am greatly indebted. I wish my limits had permitted me to publish the whole of his interesting document; but I have inserted the most essential parts of it, in detached places, where I thought they would be most effective, and I beg to offer my best thanks for them.

A word or two I should add about the language I have put into the mouths of the hillmen. It is neither the Highland nor Lowland dialect, but such, I believe, as is spoken in Perthshire. The English, which the natives of this country have, is daily improving by their intercourse with sportsmen and their followers from the South, and they now intermix their sentences with many words spoken as correctly as they are in any part of England.

The superstitions and traditions which form some portion of the following pages, being current in the country, have probably found their way into other publications; of this I know nothing—it may or may not be so—I can only say that I had them from the best authorities, and from the fountain-head. It has come, however, to my knowledge, since I have sent these pages to the press, that the trial of Duncan Terig has been mentioned in Sir W. Scott's Demonology. Had I known this before, I should not perhaps have dwelt so long upon the story, interesting as it is.

As to the graphic illustrations of the sport, I am happy to say that I have had the benefit of the talents of two most eminent gentlemen of the same family; the frontispiece and vignette are from the celebrated hand of Edwin Landseer. The figures and animals in the lithographs are, with one exception only, drawn by Charles Landseer, author of "The Parting Benediction," and other well-known splendid works. The exception is the plate which represents the "Looking for a Wounded Deer," for the whole of which, as well as for the landscape part, in every subject introduced, the author alone is responsible. They are not correct views, but only general recollections of the forest scenery. None of the figures are intended for portraits.

CONTENTS.

CHAPTER I.

Descriptive character of the red deer—Royal harts—Shedding and renewal of the horns—Weight of deer—Donald MacKay's deer-trap—Rutting season—Combats of stags—Deer stalked while fighting—Calving of hinds—Shyness and defensive instincts of deer—The bay—Traditional longevity—Red deer venison—Sir Walter Scott's Letter—Singular instance of a stag's ferocity—Deer-drive in Atholl in 1563—Hunting the stag—Deer-stalking, ... 25

CHAPTER II.

Start from Blair Castle—Bruar Lodge—A comrade joins—Ascent of Ben Dairg—Ptarmigans—Forest scenery—Spirit-stirring interest—A hart discovered—Manœuvring—Wading a burn—Getting a quiet shot—Dogs slipped—The bay in a mountain cataract—Dogs in peril—Death and gralloching of the deer—Cruel death of a deer-hound—Origin and antiseptic property of peat bogs—Ascent of Ben-y-venie—A herd discovered—Plan and manœuvring—Alarm and movements of the deer—An injudicious shot—A successful one—A deer-hound slipped—Bay—Strange adventure—A wild huntsman Encounter with a bear—Loss of a huge salmon—The Gown-cromb of Badenoch and his story, ... 58

CHAPTER III.

Forests of Badenoch, their rights and divisions—Legend of Prince Charles—Cluny Macpherson—Adventure with a wolf—Macpherson of Braekally—Children lost on a moor—Sportsmen benighted—Witchcraft—Uncomfortable position—Fraser's cairn—Boundaries of Gawick—Fate of Walter Cumming—Wrath of a fairy—Destructive avalanche—Convivial resolution—Arrival at Bruar Lodge during the night-storm, . 99

CHAPTER IV.

Necessary qualifications for a deer-stalker—Curious attitudes required—Sleep almost superfluous—Advantages of baldness—Self-possession indispensable—Abstinence from drinking, and restrictions in food—Gormandiser's pastime—Royal diversion—Sportsman's philosophy—George Ritchie the fiddler—Crafty movements—Currents of air—Passing difficult ground—Range of the rifle—Firing at the target—Tempestuous winds—A tyro's distress—Overwhelming kindness—Of speed and wind—John Selwyn—Wilson the historian—Glengary, ... 112

CHAPTER V.

A Scotch mist—Visions of auld lang syne—Retrospect—The mist clears—How to carry the spare rifles—Storm in the mountains—Sportsmen struck by a thunderbolt—Willie Robertson's lament—Macintyre's death—Deer seen on the move—Vamped up courage—Making a dash—Unexpected success—Dogs fighting, ... 130

CHAPTER VI.

The forest of Atholl—Probable number of deer, and their size—Cumyn's cairn—Highland vengeance—Fatal accident—Principal glens—Glen Tilt—Marble quarries—Roe deer—Lakes and lodges—Merry foresters—Forest song—Cuirn Marnick—Last execution at Blair—Arrest of a murderer—Royal feasting and hunting—Palace in the forest, and Highland cheer—Burning of the palace—Kilmavonaig beer—Cumming's death—Belief in witchcraft—M. G. Lewis's legendary tale of the Witch of Ben-y-gloe, ... 142

CHAPTER VII.

Deer drive to Glen Tilt—Anticipated sport—The deer-stalker's rhymes—The start from Bruar Lodge—Combat of stags—Cautious exploring—Stalking the great Braemar hart—The shot and bay—Preparation for driving the deer—Dalnacardoc chamois—A French sportsman—The ambuscade, skirmish, and slaughter—Shot at the black deer—The party assembled The last hart brought to bay—The bay broken—The death-shot—A carpet knight—Condoling with a victim—The Count's adventure—Chase and capture of a poacher—A quiet shot—Granting a favour—Termination of the day's sport, . . 166

CHAPTER VIII.

Forest contracts—Wandering poachers—English vagabonds—Adventure at Felaar—Highland vampire—Peter Breck's backsliding—Trap baited with whiskey—The Gaig pet stolen—Poacher's adventure—Desolate situation—A Highland witch—Chisholm's cave—Freebooter's life—John More—Sutherland monster—A priest in jeopardy—Highland Robin Hood—Our-na-kelig—The widow's hospitality—Rival

poachers in Atholl—Adventure in Glen Tilt—Rob Doun—
Curious trial for murder—A polyglot ghost—Ghost no
lawyer, ... 191

CHAPTER IX.

Broad awake—Arrangements for the day—A ticklish point—
Serpentine movements—Disappointment—White kid gloves
—Contest of skill—Escape of the deer—Good sport—Close
combat—A ride on a stag—Remarkable prowess—Contest
with a phoca—The drive begins—Shots and untoward acci-
dent—Corrie's sagacity and night watch—The coup d'essai—
Past deeds—Eagles killed by a boy—Driving the herd—
Legend of Fraser's cairn—The Lord of Lovat's Raid—Strong
taint of deer—Nervous excitement—Ambuscade at the wood
—Noble sport—The old Blair pony—Return to the castle, . 215

CHAPTER X.

Original Scotch greyhound—Fingal and his retinue—Bran and
Phorp—Their death—The lurcher—Glengarry's dogs—Of
blooding deer-hounds—Four-footed Hannibal—Sir William
St. Clair's dogs, ... 241

CHAPTER XI.

Occupation of Forest Lodge—Autumnal blasts—Sullen fuel—
The sport begins—Deer-stalker distressed—A sharp walk—
Lying in ambush—The fatal spot reached—Herd in jeopardy
—Peter Fraser's humanity—His penmanship—The lament
—The moors, ... 249

CHAPTER XII.

Dogs of ancient Britain—Irish dogs sent to Rome—Early Scottish dogs—Sculptured stones at Meigle—the Miol-chu—The
mastiff and greyhound—Recreation of Queen Elizabeth—
Dogs of Epirus—Irish wolf-dog—Proportions of a deer-hound
—Failure of crosses in breeding—Deer dogs of Colonsay,
and dimensions of Buskar—Expedition from Colonsay—
Cavern scene—Wild scenery in Jura—Stag discovered—
Stalking him—The start and course—His death—Speed and
bottom of deer-hounds—Decay of the ancient race, . . 260

FORESTS OF SCOTLAND.

The Sutherland Forests.—Dirrie-Chatt and its boundaries—
Forest of Dirrie-more, its character and limits—Number of
deer—Deer dykes—Wolves in Sutherland—Death of the last
wolf—Traditions of Fingal—Slaughter of a wild boar—
Dermid and Grana—Angus Baillie—The humble garron, . 279

a

CONTENTS.

Forests and Deer-haunts in Ross-shire.—Gairloch—Balnagown Forest—Easter Ross, Calrossie, and Coigach—Isles of Harris and Lewis, ... 290

Account of Coul.—Uncouth fire-arms, ... 293

Forest of Applecross.—The Laird's sport, ... 295

Forest of Glengarry.—Sagacity of a blood-hound—Wild work, . 297

The Duke of Gordon's Deer Forests.—Glenfeshie—Gawick—Glenfiddich—Glenmore, etc., ... 298

Forest of Invercauld, formerly royal—Weight of deer, etc., . 300

Forest of Marr.—Wild boar and rein-deer—Battue of the olden time, ... 304

Forest of Corrichibah.—Number and condition of deer—Mode of killing them—Translations from the Gaelic poetry of Duncan Macintyre—Spring in Bendouran—Lament for the dell of mist, ... 307

Forest of Glenartney.—Boundaries—Weight of deer, etc., . 312

The Forest of Jura.—Description of Tarbet—Deer crossing to Islay, etc., ... 313

The Isle of Skye and North Uish.—Number of deer—Method of killing them, etc., ... 316

Loch Etive and Dalness.—Tradition concerning a white hind—adventure and disastrous death of a poacher, etc., . . 316

APPENDIX.

Highest Hills in the Forest of Atholl, ... 318

Evidence relating to the Trial of Duncan Terig, *alias* Clerk, and Alexander Bain Macdonald, for the Murder of Sergeant Davies, ... 320

LIST OF ILLUSTRATIONS.

1. Frontispiece—Fighting Harts—A Forest Joust. By EDWIN LANDSEER, ... } *To precede the printed Title.*
2. Title-Vignette—Group of Dogs—Buscar, a Highland Deer-hound, of the *original* breed, belonging to Mr. MacNeill: a Fox-hound, Blood-hound, and Greyhound—from crosses of which the *modern* Deer-hound is obtained—and a Terrier. By EDWIN LANDSEER, ... }

3. Getting a Quiet Shot, ...	*To face page*	66
4. Deer at bay in a Torrent, ...	,,	70
5. Looking for a Wounded Deer, . . .	,,	90
6. Left behind in a Dubious Position, . .	,,	126
7. Lifting the Deer out of a Burn, . . .	,,	142
8. The Witch of Ben-y-gloe, ...	,,	162
9. East from Blair Castle, ...	,,	166
10. Coming in for a Shot, ...	,,	174
11. Shots from Cairn-cherie, ...	,,	224
12. Preparing the Deer for being left on the Moor,	,,	256

"My heart's in the Highlands, my heart is not here;
My heart's in the Highlands a hunting the deer;
A chasing the wild deer, and following the roe,
My heart's in the Highlands wherever I go."
—*Old Song.*

DAYS OF DEER-STALKING.

DAYS OF DEER-STALKING.

CHAPTER I.

OF THE NATURE AND HABITS OF THE RED DEER.

Descriptive character of the red deer.—Royal harts.—Shedding and renewal of the horns.—Weight of deer.—Donald M'Kay's deer-trap.—Rutting season.—Combats of stags.—Deer stalked while fighting.—Calving of hinds.—Shyness and defensive instincts of deer.—The bay.—Traditional longevity.—Red deer venison.—Sir W. Scott's letter.—Singular instance of a stag's ferocity.—Deer-drive in Atholl in 1563.—Hunting the stag.—Deer-stalking.

> I am a hart by Greekes surnamed so,
> Because my head doth with their tearmes agree;
> For stately shape few such on earth do goe,
> So that by right they have so termed me.
> For king's delight it seems I was ordayned,
> Whose huntsmen yet pursue me day by day,
> In forrest, chace, and parke, I am constrained
> Before their hounds to wander many a way.
> Wherefore who lyst to learne the perfect trade
> Of venerie, and therewith all would know
> What properties and virtues nature made
> In me poor hart (O harmless hart!) to grow,
> Let him give ear to skilfull Trystram's lore
> To Phœbus, Fowylloux, and many more.*

"*Cervus Elaphus, cornibus ramosis, teretibus, recurvatis.*"
—Linn. Eight cutting teeth in the upper jaw, and none in the lower.

Stags are found in all the northern regions, Lapland, perhaps, excepted; in Asia, especially in Tartary, and in the northern provinces of China; they are also found in America. Those of Canada differ from ours only in the length of their horns, and direction of their antlers, which are not so straight as with us, but are turned backward, so that the end of each points to the stem of the horns.

---

* The noble Art of Venerie, translated from the French, p. 39.

The colour varies slightly, but is usually of a reddish brown, nearly black about the face, mingled with grey; a dark list down the hind part of the neck and between the shoulders, and a light sort of buff colour between the haunches and underneath.

The horns vary in size and number of branches; partly owing to the age of the animal, and partly from other causes; and it must be remarked, that deer with few points to their horns are sometimes larger and fatter than those with many branches. In the forest of Atholl we had no technical names for harts of different ages; but they are thus distinguished by park-keepers, and by those gentlemen who keep stag-hounds in England:—

Before deer are one year old they are called (male and female) *Calves;** after one year old the male is termed a *Brocket;* at three, a *Spire;* at four, a *Staggart;* at five, a *Stag;* and at six a *warrantable Stag.* He may afterwards be called a Hart. The female, after one year old, is termed a *Hearst;* and at three years old a *young Hind.*

The female does not cohabit with the male till three years old. She never has more than one calf at a time, though the contrary opinion has been entertained.

The stag's brow bay and tray antlers are termed his *Rights;* the upright points on the top of his horns are called *Crockets;* the horn itself the *Beam;* the width the *Span;* the rough part of the base the *Pearls.*†

A *Brocket* has only knobbers, and small brow antlers; a *Spire,* brow and uprights; a *Staggart,* brow, tray, and uprights; a five-year old, brow, bay, and tray; two on top, that is, a crocket on one horn, and an upright on the other. A *warrantable Stag* has brow, bay, and tray, and two points on the top of both horns. After this age their heads vary very much in appearance.

---

* Some limit the term of calf to six months only.
† I am aware that these terms do not exactly correspond with those mentioned in all the old authorities, neither do the latter always accord with each other. I have taken my nomenclature from the Devonshire Hunt, as the best authority. It has been founded considerably above a century. Wriothesly, second Duke of Bedford, is the first person to whom it can be traced: he died at Tavistock, in 1711. There are about 313 deer in all the covers. Seventy were killed by the late Lord Graves in two seasons.

If the impression of a deer's foot measures full two inches at the heel, he is warrantable; if three inches, and the hoofs mark deeply in the ground, allowing for its nature, he is a large, heavy, old deer. Such bring up their hind feet to the impression made by their fore ones.

The tread of a hind is much narrower and longer than that of the male, particularly at the toe, whilst the hart's is broad and round at that point, instead of being narrow.

> "Then, if he ask, what slot or view I found,
> I say the slot or view was long on ground ;
> The toes were great, the joynt bones round and short,
> The shinne bones large, the dew-claws close in port :
> Short joynted was he, hollow-footed eke,
> An hart to hunt as any man can seeke."—*Art of Venerie.*

The mark of a deer's tread is called his *slot ;* his haunt is termed his *lair ;* where he lies down, his *harbour* or *bed ;* where he rolls himself, his *soiling pool ;* his breaking place over a hedge, his *rack ;* when he goes to water it is called going *to soil ;* if headed back, it is called *blanched ;* if he stops in a river, or lies down in a pool, during the chase, it is called *sinking himself.*

Harts that are crowned with three points at the upper extremity of each horn are termed *royal.*

We read, also, of the *hart royal proclaimed.* Manwood mentions a fact, which he found on record in the Castle of Nottingham : it is dated in the time of Richard I., who, having roused a hart in the forest of Sherwood, pursued him as far as Barnsdale in Yorkshire, where the animal foiled and escaped his hounds. The king, in gratitude for the diversion he had received, ordered him to be immediately proclaimed at Tickill, and at all the neighbouring towns, the purport of which was to forbid any one to molest him, that he might have free liberty to return to his forest.

"Some gentlemen, in the time of Henry III., having destroyed a white hart, which had given the king much diversion (and which had probably been proclaimed), his majesty laid a heavy fine upon their lands, an acknowledgment of which was paid into the exchequer so late as the reign of Queen Elizabeth."*

---

* Cam. Brit. p. 59.

Hutchins, in his History of Dorsetshire, says, "It is paid to this day."*

Deer shed their horns annually: the oldest harts shed them first, about the beginning of April; the younger ones follow in succession, according to their age and condition. The new horns attain their full growth in three months, and appear about ten days after the old ones are shed. It is not very long since a hart fell under the close observance of a forester, whilst in the act of shedding his horns, in a forest in Sutherland. Whilst he was browsing, one of his antlers was seen to incline leisurely to one side, and immediately to fall down to the ground: the stag tossed up his head, as if in surprise, and began to shake it pretty violently, when the remaining antler was discarded also, and fell some little distance from him. Relieved from this weight, he expressed his sense of buoyancy by bounding high from the ground, as if in sport, and then, tossing his bare head, dashed right away in a confused and rapid manner.

The shedding of the horns continues till the beginning of June; but deer of a year old will carry them till August or September: these new horns are very sensitive, and the harts at this time avoid bringing them into collision with any substance. When they fight, they rear themselves upon their hind legs, and spar with their fore feet, keeping back their heads. They carry their horns just as long as the hind carries her fawn, which is eight months. They are not always shed at the same time, but one of them occasionally drops a day or two after the other. I myself have seldom found any other than single horns in the mosses of the forest. It is a remarkable fact, however, that the number which are picked up in any forest bears no proportion to those which are shed; and this cannot arise from their being overlooked, for they are a valuable perquisite to the keepers, and there is no part of the forest that is not traversed by them in the course of the season.

What, then, becomes of them? Hinds have been seen to eat them: one will consume a part, and, when she drops it, it will be taken up and gnawed by the others. The late

---

* Vol. ii. p. 492.

Duke of Atholl, indeed, once found a dead hind which had been choked by a part of the horn, that remained sticking in its throat. It is not, however, credible that all those which are missing are disposed of in this way; they rather seem to be thus eaten from wantonness and caprice,—and I am not able to account satisfactorily for their disappearance.

The new horns which deer acquire annually are covered with a thick sort of leaden-coloured skin, which remains on them till the deer are in good condition: it then begins to fall off, and, for a short space, hangs in shreds, ragged and broken; but they remove it as quickly as they can, by raking their antlers in the roots of the heather, or in such branches of shrubs as they can find adapted to the purpose. When they have shaken off this skin, which is called the velvet, and which disappears in the months of August and September, they are said to have clean horns; and, as these deer are in the best condition, they are the particular object of the sportsman.

If a hart is cut when a fawn, he will never have horns; and if he is cut when five or six years old, after his horns have attained their full growth, he will never drop them; and, if he be cut when he has dropped them, they will never be renewed. This is mentioned in Buffon, and has been confirmed to me by Mr. John Crerer, who is a close observer of nature, and has had sixty years' experience in the forest of Atholl. But I once killed a very large fat hart on the top of Ben Dairg, in the month of September, which had not been cut, and still had no horns at all.

I myself have often observed, that if a hart has one of his horns ill grown, and inferior to the other, he will, upon examination, be found to have a gun-shot, or some other bad wound, on the side where the horn is faulty.

Many horns of the Cervus Elaphus have been found in peat bogs and shell marl; and, as these have the os frontis attached to them, they could not have been cast in the ordinary way; but must either have belonged to deer that died of old age or disease, or to such as might have been mired in endeavouring to land, where the bottom was soft and quaggy. Many, probably, have perished in this way, as the horns are generally found in an upright position.

A vast quantity of these horns, and, indeed, whole skeletons of deer, have been found, within this last century, in the small lakes of Forfarshire.* Indeed, antlers and skeletons of full grown stags are amongst the most common remains of animals in peat. Horns so found are infinitely larger than any which I have ever seen on living animals of the same species.

It must be inferred, therefore, that the animals themselves were likewise of very superior dimensions. At first sight this seems difficult to account for; but when we take into consideration the altered circumstances of the country,—that immense tracts of wood have given place to barren bogs, in the manner explained by Mr. Lyell, and mentioned in the course of these pages, and that the deer have thus been limited in food and shelter,—we can no longer be at a loss to account for this degeneracy.

The red deer is not a very hardy animal: he does not by choice subsist on coarse food, but eats close, like a sheep. With his body weakened and wasted during the rutting season in the autumn, exposed to constant anxiety and irritation, engaged in continual combats, he feels all the rigours of winter approaching before he has time to recruit his strength:—the snow-storm comes on, and the bitter blast drives him from the mountains. Subdued by hunger, he wanders to the solitary sheelings of the shepherds; and will sometimes follow them through the snow, with irresolute steps, as they are carrying the provender to the sheep. He falls, perhaps, into moss pits and mountain tarns, whilst in quest of decayed water plants, where he perishes prematurely from utter inability to extricate himself. Many, again, who escape starvation, feed too greedily on coarse herbage at the first approach of open weather, which produces a murrain amongst them, not unlike the rot in sheep, of which they frequently die. Thus, natural causes, inseparable from the condition of deer in a northern climate, and on a churlish soil, unsheltered by woods, conspire to reduce these animals to so feeble a state, that the short summer which follows is wholly insufficient to bring them to the size they are capable of attaining under better management.

* Vide Lyell's Geology, vol. ii. p. 259.

If we look at the difference in size and weight of two three-year-old beasts, the one belonging to a good, and the other to a bad farmer, we shall find that difference to amount to nearly double. The first animal is well fed for the sake of the calf, both in winter and summer; and the last, from insufficient keep, loses in winter what it has gained in summer, and requires double the food in the succeeding season to restore it to what it was at the commencement of winter. Thus it is with the deer.

As a proof of this position, I may mention, that such stags as have, for the most part, abandoned the Scotch mountains, and pastured in the large woods in the low country, have been found considerably to exceed the hill stags in size and condition. The late Duke of Atholl killed a hart that had been feeding for four seasons in the woods of Dunkeld, where he remained, with twelve others, during nine months of the year. He weighed thirty stone six pounds imperial as he stood. His horns weighed thirteen pounds two ounces; but they were still inferior to such as have been found buried in peat mosses. The fat on his haunches was four inches and one-eighth thick, though he was killed in July, much too early in the season to have arrived at his full condition.

In the year 1836, an outlying stag was killed at Woburn, which weighed thirty-four stone imperial as he stood. These are much higher weights than are to be found in the forest of Atholl.

In the forest of Glengarry, where the snow never lies long, where there is much rich pasture in the low grounds, sweet grass on the hill-tops, and large woods for shelter, the late Glengarry killed a hart, which weighed twenty-six stone after the gralloch or offal was taken out: now, allowing six stone six pounds for the gralloch (computing it at about one-fourth of the entire weight), this noble animal must have been thirty-two stone six pounds as he stood.

From the accounts that have been sent to me from the various forests in Scotland, I am inclined to think that the average weight of the best deer in Sutherland is superior to that of the other forests. It reaches about fifteen stone, Dutch, sinking the offal; and stags are occasionally killed of seventeen stone; and, in the forest of Ben Hope, of a somewhat larger size. Now, Dutch weight reckons sixteen pounds to the stone, and seventeen ounces and a half to the pound; so that adding the offal, and reducing the whole to imperial weight, a stag of fifteen stone Dutch would be about twenty-five stone imperial as he stands.

In corroboration of what has been advanced above, as to the starving condition of the Highland deer in severe weather, I shall mention a fact that happened about the end of the last century.

One Donald M'Kay, a farmer, who lived in a remote glen on the estate of Reay, in Sutherland, received so much injury from the depredations of the forest deer, which made continual inroads upon his crops, invading him from the west and from the north, that he at length marched off to Tongue, the residence of his landlord and chief, to endeavour to procure some redress. Having obtained an audience, Lord Reay, who probably gave little credit to his tale, told him to go back and pound the deer whenever they trespassed in future. Donald did not presume to say aught against his reception, though he was bitterly vexed at having walked forty mortal miles for nothing.

On his arrival at his little farm, he set his wits to work to devise some plan for making use of the permission which had been conceded to him. Donald was a shrewd fellow; but it was not particularly easy to pound the denizens of the mountains. He was pretty secure for the present, as he had built a large barn, and kept his crop on rafters, out of the reach of all depredators; when the winter came on, he put part of this crop very carefully into one end of his barn, and barred it in with sticks and fir roots, in such a manner that no beast or person could get at it.

About the end of November a very heavy fall of snow came on, and the ground was wholly covered with it. The second or third night after the storm fell the wind was from the west; and Donald spread the sheaves on the rafters, the barn door giving eastward: he then threw the door wide open, and tied a long rope of hair to it, the end of which he took in at the only window that was behind the dwelling-house. Well did he know that the storm would drive the deer to his house in the still hour of night to search for the least particle of such fodder as might be dropped betwixt the barn and the byre in feeding the bestial. He therefore took his station within the window, with the end of the tether in his hand. He had not been long in this situation before he saw the gaunt and starving animals approaching. They came forward slowly and cautiously, stopping at intervals, and examining every object; at length the cravings of nature prevailed, and two hinds walked into the barn, and began eating the corn. The stags soon followed; and some of them had great difficulty in getting their antlers through the narrow door.

As soon as ten deer had fairly entered, Donald pulled the tether, and made the door secure. More blithe than before, he set off a second time for Tongue, travelling as fast as his legs could carry him. On his arrival, he craved an audience of Lord Reay, and told him in Gaelic that he had followed his advice, and pounded ten of his deer. "I might," said he, "as well have had a hundred as ten; but I could not afford to give them straw whilst I came to report the affair to your lordship."

Not a little incredulous, Lord Reay despatched two men to ascertain the truth of the matter. The deer were found imprisoned as related, and were liberated. Donald M'Kay then came to terms with his chief, who very handsomely gave him his little farm rent-free for his life, upon condition that he would not pound his deer for the future.

It is remarkable for how short a time deer continue in season in the cold climate of the north, owing to the backward vegetation, and the causes already alluded to. In warmer climates they come in sooner; and we are informed by Aristotle that, in Greece, the rutting season commenced in the beginning of August, and terminated about the end of September.

In Scotland this season varies slightly according to the weather; if mild and warm, the deer do not rut so soon, but, if the weather is cold and frosty, the harts are brought forward earlier; indeed, it is quite surprising what a few cold nights will effect in this way.

About the end of September, and the first week in

October, the harts swell in their necks, have a ruff of long wiry hair about them, and are drawn up in their bodies like greyhounds. They now roll restlessly in the peat pools till they become almost black with mire, and feed chiefly on a light coloured moss, that grows on the round tops of the hills, so that they do not differ so entirely from the rein-deer in their food as some naturalists have imagined.

In this state of rutting they are rank, and wholly unfit for the table. Such deer a good sportsman never fires at; but many may be found at this time, not so forward, but perfectly good; and they are, of course, easily distinguished. This is a very wild and picturesque season. The harts are heard roaring all over the forest, and are engaged in savage conflicts with each other, which sometimes terminate fatally. When a master hart has collected a number of hinds, another will endeavour to take them from him: they fight, till one of them, feeling himself worsted, will run in circles round the hinds, being unwilling to leave them: the other pursues; and when he touches the fugitive with the points of his horns, the animal, thus gored, either bounds suddenly on one side, and then turns and faces him, or will dash off to the right or the left, and at once give up the contest. The conflict, however, generally continues a considerable time, and nothing can be more entertaining than to witness, as I have often done, the varied success and address of the combatants. It is a sort of wild just, in the presence of the dames who, as of old, bestowed their favours on the most valiant. Sometimes it is a combat *à l'outrance*, but it often terminates with the effect of the horn of Astolfo.

In solitary encounters, there being no hinds to take the alarm, the harts are so occupied, and possessed with such fury, that they may be occasionally approached in a manner that it would be vain to attempt at any other time. From the summit of a mountain in Atholl forest, I once saw two harts in fierce contention, in a mossy part lower down the hill. I came into sight at once, not expecting to see deer in the situation in which these happened to be. I could neither advance straight forward, nor retreat, without danger of giving the alarm. One possibility alone was open to me; this was to get into the glen to their right when I should be entirely hidden from their view, and then come up, concealed by the hill, as nearly opposite to them as possible. I was certainly a very considerable distance to the north of them; but my position was so bad that I looked upon my chance as a mere nothing. I lay down, however, flat on my back, amongst the rugged and loose stones of Cairn-marnac, with a rifle in my hand; Thomas Jamieson, with the other rifles, placed himself behind me in the same comfortable position. We had a full view of the deer for some time, so that with their ordinary vigilance they would undoubtedly have seen us; the stones, however, formed an uneven outline, which was in our favour, and thus we did not absolutely attract their notice. Whilst the stags were fiercely engaged, we worked our way down on our backs, looking askance; when they rested for a space, and sometimes they would do so on their knees, from mere exhaustion, we moved not a limb: and in this manner we wormed ourselves gradually into the glen, not without certain uncomfortable bruises. Then, being out of sight we sprang up, and made the best of our way to the point immediately below them; and moving cautiously up the hill, which was sufficiently steep for our purpose, we came all at once in full view of one of the combatants, who was then alone; he sprang off at full speed, but all too late for his escape, for my ball struck him dead on the spot. His antagonist, I imagine, had been beaten off. I expected to have killed them both.

A conflict of this savage nature, which happened in one of the Duke of Gordon's forests, was fatal to both of the combatants. Two large harts, after a furious and deadly thrust, had entangled their horns so firmly together that they were inextricable, and the victor remained with the vanquished. In this situation they were discovered by the forester, who killed the survivor, whilst he was yet struggling to release himself from his dead antagonist. The horns remain at Gordon Castle, still locked together as they were found. Mezentius himself never attached the dead body to the living one in a firmer manner.

A hart will rut for about a week, after which period he

becomes weak and exhausted, and seeks some solitary spot where he may recruit himself in peace: no wonder, indeed, since during this week he is constantly with a large herd of hinds, at times fighting, and always in a state of the highest irritation; for, at the darkening, another and another hart will come in, and take some of the hinds from him; he then gives chase furiously, but is obliged to return after a short space for fear of losing the remainder. At length the old harts that have rutted, collect and go off together in large parcels, and the younger ones succeed to the hinds. During the winter they have long wiry coats of a lighter cast of colour, and are wholly without fat, and in every respect unfit for the table. The winter coat begins to come off when they drop their horns, and the new coat appears about the middle of June.

Neither Mr. John Crerer, who has followed deer in the forest of Atholl for sixty years, nor any other individual there, has ever seen a hart cover a hind.

The period of gestation in a hind is eight months. She drops her fawn in high heather, where she leaves it concealed the whole of the day, and returns to it late in the evening, when she apprehends no disturbance. She makes it lie down by a pressure of her nose; and it will never stir or lift up its head the whole of the day, unless you come right upon it, as I have often done. It lies like a dog with its nose to its tail. The hind, however, although she separates herself from the young fawn, does not lose sight of its welfare, but remains at a distance to the windward, and goes to its succour in case of an attack of the wild cat or fox, or any other powerful vermin. I have heard Mr. John Crerer say, and it is doubtless true, that if you find a young fawn that has never followed its dam, and take it up and rub its back, and put your fingers in its mouth, it will follow you home for several miles; but if it has once followed its dam for ever so small a space before you found it, it will never follow human beings. When once caught, these fawns or calves are easily made tame; and there were generally a few brought up every year by the dairymaid at Blair. I speak of hinds only; stags soon turn vicious and unmanageable. When the calf is old enough to keep up with a herd of deer, and to take pretty good care of itself, its mother takes it off, and leads it into ground that can be travelled without difficulty, avoiding precipitous and rocky places.

Hinds that have calves have no fat whatever upon them, and are fit only for soup, or perhaps for stewing. A good sportsman will never fire at them: indeed, it is reckoned a disgrace to do so; and a most wanton act of cruelty it certainly is. The best shots, however, will occasionally kill them accidentally; for they come up so rapidly before the hart you are aiming at, that they often, like Polonius, get that which was meant for their betters. Those hinds, however, that have not bred for the season, are lawful game; they are killed late in the year, and their venison is fat and more delicate than that of the harts. They are called Yell or Yeld Hinds, these terms signifying *barren*. They are known by their sleek and compact make; but it requires a very experienced eye to distinguish them at a distance with certainty; and I must confess I have sometimes been egregiously deceived. They come into season when the harts go out.

Deer, except in certain embarrassed situations, always run up wind; and so strongly is this instinct implanted in them, that if you catch a calf, be it ever so young, and turn it down wind, it will immediately face round and go in the opposite direction. Thus they go forward over hill-tops and unexplored ground in perfect security, for they can smell the taint in the air at an almost incredible distance. On this account they are fond of lying in open corries, where the swells of winds come occasionally from all quarters.

I have said that deer go up wind; but, by clever management, and employing men to give them their wind (those men being concealed from their view), they may be driven down it; and in certain cases they may easily be sent, by a side wind, towards that part of the forest which they consider as their sanctuary.

It is to be noted, that on the hill-side the largest harts lie at the bottom of the parcel, and the smaller ones above; indeed, these fine fellows seem to think themselves privileged to enjoy their ease, and impose the duty of keeping guard upon the hinds and upon their juniors. In the performance of this task, the hinds are always the most vigilant, and when deer are driven, they almost always take the lead. When, however, the herd is strongly beset on all sides, and great boldness and decision are required, you shall see the master hart come forward courageously, like a great leader as he is, and, with his confiding band, force his way through all obstacles. In ordinary cases, however, he is of a most ungallant and selfish disposition; for, when he apprehends danger from the rifle, he will rake away the hinds with his horns, and get in the midst of them, keeping his antlers as low as possible.

There is no animal more shy or solitary by nature than the red deer. He takes the note of alarm from every living thing on the moor,—all seem to be his sentinels. The sudden start of any animal, the springing of a moor-fowl, the complaining note of a plover, or of the smallest bird in distress, will set him off in an instant. He is always most timid when he does not see his adversary, for then he suspects an ambush. If, on the contrary, he has him full in view, he is as cool and circumspect as possible; he then watches him most acutely, endeavours to discover his intention, and takes the best possible mode to defeat it. In this case he is never in a hurry or confused, but repeatedly stops and watches his disturber's motions; and when at length he does take his measure, it is a most decisive one: a whole herd will sometimes force their way at the very point where the drivers are the most numerous, and where there are no rifles; so that I have seen the hill-men fling their sticks at them, while they have raced away without a shot being fired.

When a stag is closely pursued by dogs, and feels that he cannot escape from them, he flies to the best position he can, and defends himself to the last extremity. This is called, going to bay. If he is badly wounded, or very much over-matched in speed, he has little choice of ground; but if he finds himself stout in the chase, and is pursued in his native mountains, he will select the most defensible spot he has it in his power to reach; and woe be unto the dog that approaches him rashly. His instinct always leads him to the rivers, where his long legs give him a great advantage over the deer-hounds. Firmly he holds his position, whilst they swim powerless about him; and would die from cold and fatigue before they could make the least impression on him. Sometimes he will stand upon a rock in the midst of the river, making a most majestic appearance; and in this case it will always be found that the spot on which he stands is not approachable on his rear. In this situation he takes such a sweep with his antlers that he could exterminate a whole pack of the most powerful lurchers, that were pressing too closely upon him in front. He is secure from all but man; and the rifle-shot must end him. Superior dogs may pull him down when running, but not when he stands at bay.

It is worthy of remark, that when a cold hart (meaning one that has not been wounded) takes the bay, and it is broken afterwards by an unskilful approach, or by any other means, the same dog or dogs which outran it at first, seldom succeed in bringing it to bay a second time. The dogs exhaust themselves with their clamour and exertions, whilst the hart is in a comparative state of rest, and recovers his wind.

There is an opinion amongst many, founded upon tradition, that the deer attains to a very extraordinary old age, amounting to some hundreds of years: "*Longa et cervina senectus*," saith Juvenal. But the ground and authority of this conceit, according to Sir Thomas Browne, " was first hieroglyphical, the Egyptians expressing longevity by this animal; but they often erected such emblems upon uncertainties, and convincible falsities; for Aristotle, first, and Pliny long after, declared, that the Egyptians could make but weak observations on this matter; for although it was said that Æneas feasted his followers with venison, yet Aristotle affirms that neither deer or boar were to be found in Africa: and how far they misconstrued the lives and duration of animals is evident, from their conceit of the crow, which they presume to live for five hundred years; and the lives of hawks, which, according to Ælian, the Egyptians reckoned at no less than seven hundred."

Setting aside the absurd story of the stag taken by Charles the Sixth, let us see if there be any modern proof that may throw light upon this subject.

In the year 1826, the late Glengarry, accompanied by Lord Fincastle, now Earl of Dunmore, was hunting in the Garth of Glengarry; the beaters had been sent into a wood, called Tor-na-carry; a fine stag soon broke forth, and was going straight to Lord Fincastle, but owing to a slight swell, or change of the current of air, he turned towards Glengarry, who fired at, and killed him.

On going up to him a mark was discovered on his left ear; the first man who arrived was asked, "What mark it was?" He replied, "That it was the mark of Ewen-mac-Jan Og." Five others gave the same answer; and after consulting together, all agreed that Ewen-mac-Jan Og had been dead 150 years, and for thirty years before his death had marked all the calves he could catch with this particular mark; so that this deer (allowing the mark to have been authentic) must have been 150 years old, and might have been 180. The horns, which are preserved by the present Glengarry, are not particularly large, but have a very wide spread.

Now this circumstance is clearly and honourably attested; it was communicated to me both by the late and present Glengarry; we must, therefore, either subscribe at once to this longevity, or we must imagine (what indeed seems to be most probable), that as the old forester's mark was evidently known to the hill-men, some of his successors might have imitated it, without the sanction or knowledge of their chief.

However this may be, it is notorious that no superstition is more prevalent amongst certain classes in the Highlands, than that which regards the longevity of deer. Hence the following adage:—

> "Tri àvis coin, àvis eich;
> Tri àvis eich, àvis duine;
> Tri àvis duine, àvis feidh;
> Tri àvis feidh, àvis firean;
> Tri àvis firean, àvis craobh dharaich."

Thus it stands in English:—

Thrice the age of a dog is that of a horse;
Thrice the age of a horse is that of a man;
Thrice the age of a man is that of a deer;
Thrice the age of a deer is that of an eagle;
Thrice the age of an eagle is that of an oak tree.

Setting aside the extravagance of this adage, I venture to mention that, according to tradition, Captain Macdonald, of Tulloch, in Lochaber,* who died in 1776, at the age of eighty-six, knew the white hind of Lochtreig for the last fifty years of his life; his father knew her an equal length of time before him, and his grandfather knew her for sixty years of his own time; and she preceded his days: these three gentlemen were all keen deer stalkers. Many of the Lochaber and Brae Rannoch men knew her also; she was purely white, without spot or blemish,—was never seen alone, and tradition furnishes no instance of any shot having been fired at the herd with which she associated.

A very large stag was known for 200 years in the Mona-lia, a range of mountains lying between Badenoch and Inverness. He was always seen alone, keeping the open plains, so that he was unapproachable. He was easily distinguished from all others by his immense proportions.

About the year 1777, Angus Macdonald, after stalking for five hours, got within shot of Damh-mor-a-Vonalia, as he was called (that is the large stag of Mona-lia); he fired, and saw distinctly with his glass that the ball had entered his left shoulder blade. He fell to the shot, but, not being severely injured, recovered, and got away.

Macdonald soon made known that he had wounded the Damh-mor, but there was some scepticism on the subject. In 1807, thirty years after this occurrence, the Damh-mor was shot four miles to the westward of the inn at Garvie-more, at the head of Badenoch. Thus it was:—

John Macdonald (innkeeper there, and brother to Angus, who wounded the deer as above), having heard that the hart was seen in his country, went in quest of him; and after stalking nearly a whole day in August, got within distance, and brought him down. After a minute examination, the ball of 1777 was found in the left shoulder, an

---

* Communicated by Mr. Macgregor.

inch under the skin, which still retained the mark of an old-standing perforation. The horns were by no means remarkable in point of size; but that on the left, being the side on which the deer was wounded, was ill-shaped, and defective.

The belief in the extraordinary longevity of the red deer is not peculiar to the Highlands. I have been informed by a gentleman, who has frequently attended the Duke of Sage Coburg's deer-hunts, that he has very lately seen in the mountains of Thuringia a stag of stupendous height and dimensions, whose great age is quite a tradition, having been handed down from father to son in the village from a very remote and untraceable period of time; though he still appears in full vigour; he has long enjoyed an indemnity, the duke having restricted every one from firing at him. The woods are of oak; and the acorns no doubt are one great cause of the large growth of the German deer.

William Twici, or Twety, grand huntsman to King Edward the Second, in his Treatise upon Hunting, mentions, amongst other beasts of the chase of the first class, the buck, the doe, the bear, the rein-deer, the elk, and the spytard; which latter, he himself informs us, is a hart of one hundred years old: these he calls beasts of sweet flight.

On the other hand, "Aristotle, drawing an argument from the increment and gestation of deer (I quote from Sir Thomas Browne), comes to the conclusion, that they are not such as afford an argument of long life: and these (saith Scaliger, his translator) are good mediums, conjunctively taken,—that is, not one without the other: for of animals, viviparous, such as live long, go long with young, and attain but slowly to their maturity and stature: so the horse, that liveth above thirty, arriveth at his stature in about six years, and remaineth above ten months in the womb; so the camel, that liveth unto fifty, goeth with young no less than ten months, and ceaseth not to grow before seven; and so the elephant, that liveth an hundred, beareth its young above a year, and arriveth unto perfection at twenty. On the contrary, the sheep and goat, which live but eight or ten years, go but five months, and attain to their perfection at two years: and the like proportion is observable in cats, hares, and conies. And so the deer that endureth the womb but eight months, and is complete at six years, from the course of nature we cannot expect to live a hundred years, nor in any proportional allowance to much more than thirty.

"Moreover, the state and declination of all animals are proportionally set out by nature; and naturally proceeding, admit of inference from each other. When long life is natural, the marks of age are late; and where they appear, the journey unto death cannot be long. Now the age of deer is best conjectured by view of the horns and teeth. From the horns there is a particular and annual account unto six years,—they arising first plain, and so successively branching; after which the judgment of their years by particular marks becomes uncertain: but when they grow old, they grow less branched, and first do lose their propugnacula or brow antlers; which Aristotle says the youngest use in fight, and the old, as needless, have them not at all. The same may be also collected from the loss of their teeth, whereof in old age they have few, or none before, in either jaw. Now these are infallible marks of age; and when they appear we must confess a declination, which notwithstanding will happen, as we are informed, between twenty and thirty."

I myself may add, that the great incitement and exhaustion during the rutting season, as well as the effort nature makes in renewing the horns annually, is an argument against longevity; and, notwithstanding the extreme respect I bear to marvellous traditions (always, I think, better attested in proportion as they are marvellous), I judge it incumbent on me to say, that the accounts I have received from park-keepers in England, where there are red deer, entirely contradict their supposed longevity.

The longest lived deer they remember in Richmond Park was the Naphill stag, turned out there by command of his majesty George the Third. Every care was taken of him, but he lived no longer than twenty years; and the present keeper, who communicated this information to me, asserted, at the same time, that the red deer in that park rarely exceed the age of eighteen years, and that their horns decrease in size after the age of twelve. The largest antlers he has met with there, with the skull part attached, weigh about twelve pounds.

The deer, like many other animals, seem to foresee every change of weather: at the approach of a storm they leave the higher hills, and descend to the low grounds, sometimes even two days before the change takes place. Again, at the approach of a thaw, they leave the low grounds and go to the mountains by a similar anticipation of change. They never perish in snow drifts, like sheep, since they do not shelter themselves in hollows, but keep the bare ground, and eat the tops of the heather.

One would imagine that in a severe storm many would perish by avalanches. But, during the long period of sixty years, Mr. John Crerer remembers but two accidents of this nature. These were in Glen Mark: eleven were killed by one fall, and twenty-one by another: the snow in its descent carried the deer along with it into the glen and across the burn, and rolled up a little way on the opposite brae, where the animals were smothered.

Harts are excellent swimmers; and will pass from island to island in quest of hinds, or change of food. It is asserted that the rear hart in swimming rests his head on the croup of the one before him; and that all follow in the same manner.

When a herd of deer are driven, they follow each other in a line; so that when they cross the stalker it is customary for him to lie quiet, and suffer the leaders to pass before he raises his rifle; if he were to fire at the first that appeared, he would probably turn the whole of them; or if he were to run forward injudiciously after a few had passed, the remainder, instead of following the others in a direct line, would not cross him except under particular circumstances and dispositions of ground, but would bear off an end, and join the others afterwards. It must be remarked, however, that when deer are hard pressed by a dog, they run in a compact mass, the tail ones endeavouring to wedge themselves into it. They will also run in this manner when pressed by drivers on the open moor. But they are sensible that they could not pass the narrow oblique paths that are trodden out by them in the precipitous and stony parts of the mountain, or encounter the many obstructions of rock, river, and precipice that rugged nature is continually opposing to them, in any other manner than in rank and file. If they did they must separate, and lose the wind, which is not their system.

They do not run well up hill when fat, but they will beat any dog in such oblique paths as I have mentioned. The hardness and sharp edges of their hoofs gives them great tenacity, and prevents their suffering from the stones; whilst a dog, having no fence against injury, is obliged to slacken his pace.

The bone also of a deer's foot is small and particularly hard; it is this peculiar construction which renders the animal as strong as he is fleet. The support and strength of the joints of the feet of all animal bodies, according to Sir E. Home, depends less upon their own ligaments than upon the action of the muscles, whose tendons pass over them. "This fact," he says, "was strongly impressed on my mind in the early part of my medical education, by seeing a deer which leaped over the highest fences, and the joints of whose feet, when examined, were as rigid in every other direction, but that of their motion, as the bone itself; but when the tendo Achilles, which passed over the joint, was divided, with a view to keep the animal from running away, the foot could readily be moved in any direction, the joint no longer having the smallest firmness."

Some old authorities attribute various medicinal virtues to certain parts of the hart; and, amongst the rest, the author of the Treatise on Venerie very gravely asserts, "That his marrow or grease is good for the gout, proceeding from a cold cause,—melting it, and rubbing the place where the pain is therewith. Also the hart first taught us to find the herb called Dictamus; for when he is stricken with an arrow or dart he seeketh out that herb, and eateth thereof; the which maketh the dart or arrow to fall out, and healeth him immediately."

Almost every part of the deer is excellent for the table: the liver, the heart, the tripe, the feet, and the white puddings, should not be neglected. The skin itself is manufactured into a soft yellow-coloured leather, which is useful for numerous purposes.

I have heard the excellence of the venison disputed by sportsmen, and others who have tasted it in the north; but I attribute this entirely to the age and condition of the sort of creature it was their lot to taste, or to the time of year in which it was killed. A hart, like most other animals, has little fat when he is growing; and if sportsmen do not distinguish, or have not the means of selection, the haunches will cut but a sorry figure at the table. But in the estimation of all the numerous guests it has been my good fortune to meet in the hospitable halls of Blair, the red deer has been infinitely preferred to the fallow; and I could name many such guests, whose judgment would be pronounced paramount in such matters. On the contrary, the haunch of the fallow deer, when brought to table at Blair, although perfect in its kind, was always neglected. There must however be a wide difference between the quality of the red deer, which are fed in English parks, and such as wander freely over the mountains, and browse on the sweet grass and heather.

I have now lying before me a letter from Sir Walter Scott, to whom I was in the habit of sending Highland venison (and who was no mean judge of the merits of a *plat de resistance*), attesting its excellence. Thus I quote from it, word for word:—

> " Thanks, dear Sir, for your venison, for finer or fatter
> Never roam'd in a forest, or smoked in a platter."

" Your superb haunch arrived in excellent time to feast a new married couple, the Douglasses, of M——, and was pronounced by far the finest that could by possibility have been seen in Teviotdale since Chevy Chase. I did not venture on the carving, being warned both by your hints, and the example of old Robert Sinclair, who used to say that he had thirty friends during a fortnight's residence at Harrowgate, and lost them all in the carving of one haunch of venison; so I put Lockhart on the duty, and, as the haunch was too large to require strict economy, he hacked and hewed it well enough."

Stags, although they have frequent and ferocious combats amongst their own species during the rutting season, have been seldom known to attack men, in any other way than in self-defence. No instance of the sort ever occurred to me, nor to Mr. John Crerer, who shot sixty years in the forest of Atholl. Once, indeed, he incurred a sort of rebuff by his own imprudence; being a very powerful man, he got behind a stag, which was at bay at Glenmore, and thought it advisable to take hold of his hind leg, and endeavour to throw him over; but when about to do so, the animal saluted him with both his hind legs, and with such effect, that one of his hoofs broke his watch, and the other struck him in the mouth, knocked out one of his teeth, and sent him sprawling on his back to the edge of the water. The only instance I ever heard of in that forest, of an offensive assault on man, was recounted to me by the late Duke of Atholl. His Grace had wounded a hart, and one of the deer-hounds flew at him and seized hold of his ear; when the duke came up, the hart sprung forward with his head down (the dog still hanging to his ear), and was rushing to the attack, but his Grace escaped the danger by sending a ball through his forehead. This, as I have said, is the only instance I ever heard of an offensive attack upon man by deer upon the wild mountains; and it must be observed, that the animal here in question was rendered furious by the dog, and by the pain of his wound. It is, however, at all times dangerous to approach a wounded deer too nearly, for, in self-defence, he would not hesitate to kill any living thing that came within reach of his pointed antlers. An instance is recorded of a red deer having beat off a tiger, which was set loose upon it in an inclosed arena, at the instance of William Duke of Cumberland. But if stags in such wild regions stand in awe of man, they have not always the same respect when they become more familiar with him.

"Some years ago," says Gilpin, "a stag in the New Forest, pressed by the hunters, and just entering a thicket, was opposed by a peasant, who foolishly, with his arms extended, attempted to turn him. The stag held his course, and darting one of his antlers into the man, carried him off some paces, sticking upon his horn: the man was immediately conveyed to Lymington, where he lay dangerously ill for some time, but at length recovered. I have heard also that when the Duke of Bedford was lord-warden of the forest, his huntsman had a horse killed under him by a stag, which he crossed in the same imprudent manner. "We read" (saith the editor of the *Noble Art of Venerie*) "of an emperor named Batels, who had done great deedes of chivalrie in his country, and yet was nevertheless slayne with a hart in breaking of a bay."

But a recent instance occurred in October, 1836, of the ferocity of a red deer when confined in a park, which, from the courtesy of the gentleman to whom it happened, I am enabled to give circumstantially.

The Hon. Mr. and Mrs. Fox Maule had left Taymouth with the intention of proceeding towards Dalguise, and in driving through that part of the grounds where the red deer were kept, they suddenly, at a turn of the road, came upon the lord of the demesne, standing in the centre of the passage, as if prepared to dispute it against all comers.

Mr. Maule being aware that it might be dangerous to trifle with him, or to endeavour to drive him away (for it was the rutting season), cautioned the postillion to go slowly, and give the animal an opportunity of moving off.

This was done, and the stag retired to a small hollow by the side of the road; on the carriage passing, however, he took offence at its too near approach, and emerged at a slow and stately pace, till he arrived nearly parallel with it; Mr. Maule then desired the lad to increase his pace, being apprehensive of a charge on the broadside.

The deer, however, had other intentions; for as soon as the carriage moved quicker, he increased his pace also, and came on the road about twelve yards ahead of it, for the purpose of crossing, as it was thought, to a lower range of the park; but to the astonishment, and no little alarm of the occupants of the carriage, he charged the offside horse, plunging his long brow antler into his chest, and otherwise cutting him.

The horse that was wounded made two violent kicks, and is supposed to have struck the stag, and then the pair instantly ran off the road; and it was owing solely to the admirable presence of mind and nerve of the postillion, that the carriage was not precipitated over the neighbouring bank. The horses were not allowed to stop till they reached the gate, although the blood was pouring from the wounded animal in a stream as thick as a man's finger. He was then taken out of the carriage, and only survived two or three hours. The stag was shortly afterwards killed.

Of the various modes practised for pursuing and killing the deer in different ages and countries, I do not profess to treat. In thinly-peopled districts, like the Wilds of North America, whose inhabitants subsist by the chase, artificial fences, stretching over vast distances, are employed to aid in driving the deer to the spots, where the pit-fall, the net, the spear, arrow, or rifle are employed for their destruction.

On the Continent, deer-driving on the grandest scale is still occasionally practised, the game of a whole province being surrounded by the marshalled peasantry of a prince or noble, and forced by the gradual narrowing of the circle to some central spot for promiscuous slaughter. Similar princely *Battues* were formerly common, when the game was more plentiful, and cultivation rarer, both in England and Scotland. As one instance among many of these, which we find recorded in the old chroniclers, and as a proof of the determined resolution of the stag when pushed to extremity, I may be permitted to quote the following account.

Spottiswood mentions in his History, "That Queen Mary took the sport of hunting the deer in the forest of Mar and Atholl, in the year 1563," of which Barclay, in his Defence of Monarchial Government, gives the following particulars:

"The Earl of Atholl prepared for her Majesty's reception by sending out about two thousand Highlanders to gather the deer from Mar, Badenoch, Murray, and Atholl, to the district he had previously appointed. It occupied the Highlanders for several weeks in driving the deer to the amount of two thousand, besides roes, does, and other game.

"The Queen, with her numerous attendants and a great concourse of the nobility, gentry, and people, were assembled at the appointed glen, and the spectacle much delighted her Majesty, particularly as she observed that such a numerous herd of deer seemed to be directed in all their motions by one stately animal among them; they all walked, stopped, or turned as he did,—they all followed him. The Queen was delighted to see all the deer so attentive to their leader, and upon her pointing it out to the Earl of Atholl, who knew the nature of the animal well, having been accustomed to it from his youth, he told her that they might all come to be frightened enough by that beautiful beast. 'For,' said he, 'should that stag in the front, which your Majesty justly admires so much, be seized with any fit of fury or of fear, and rush down from the side of the hill, where you see him stand, to this plain, then would it be necessary for every one of us to provide for the safety of your Majesty, and for our own: all the rest of those deer would infallibly come with him as thick as possibly they could, and make their way over our bodies to the mountain that is behind us.'

"This information occasioned the Queen some alarm, and what happened afterwards proved it not to be altogether without cause, for her Majesty having ordered a large fierce dog to be let loose on a wolf that appeared, the leading deer, as we may call him, was terrified at the sight of the dog, turned his back and began to fly thither whence they had come; all the other deer instantly followed.

"They were surrounded on that side by a line of Highlanders, but well did they know the power of this close phalanx of deer, and at speed; and therefore they yielded, and opposed no resistance; and the only means left of saving their lives, was to fall flat on the heath in the best posture they could, and allow the deer to run over them. This method they followed, but it did not save them from being wounded; and it was announced to the Queen that two or three men had been trampled to death.

"In this manner the deer would have all escaped, had not the huntsmen, accustomed to such events, gone after them, and with great dexterity headed and turned a detachment in the rear; against these the Queen's staghounds and those of the nobility were loosed, and a successful chase ensued. Three hundred and sixty deer were killed, five wolves, and some roes; and the Queen and her party returned to Blair delighted with the sport."

If this account by Barclay is matter of fact throughout (which I very much doubt), it would be curious to know in what manner these 2,000 men proceeded, and how they consumed several weeks in bringing down 2,000 head of deer. Such a force of men, well and equally distributed, would cover an immense tract of ground, but the wind must be changing upon them continually, and it must have required the strictest order, and perhaps fires throughout the line to keep the deer in during the dark nights, at which time they will go in any direction, either up or down wind. Even in the daytime, a cross wind might be fatal to the drive, if it were not for the enormous extent of ground that a force of 2,000 men could cover. A hundred men a mile would give less than twenty yards of interval between each man, and constitute a line of twenty miles in length. But how did all these rough-footed Highlanders subsist for two months on the barren mountains? A few days, one would think, would have been quite sufficient for their purpose. As for the number of deer that were killed, if a hundred couple of fierce and swift dogs were let loose, which we are told was not unusual, they must have pulled down a great many hinds and calves, though probably but few harts.

When the country was partially covered with wood the forests were driven, and the sportsmen occupied passes where they took their chance of sport; and this method is still occasionally resorted to in the forest of Glengarry and in other places. But, generally speaking, the system has given way to the more exciting amusement of deer-stalking.

The destruction of the woods, and the substitution of the gun for the bow and arrow and hagbute, formed quite an epoch in the habits and size of the deer, as well as in the mode of killing them.

In Sutherland, fire-arms were unknown until about the latter end of the 16th century, when a large awkward kind of blunderbuss, named by the country-people *Glasnabhean* (meaning the mountain match-lock gun), was obtained by Angus Baillie of Uppat, one of the most noted of the Sutherland foresters of whom we have any correct account; and it was used by him with great effect in some of the conflicts and skirmishes that were of frequent occurrence in those days.

This memorable appearance of *Glasnabhean* * took place in the year 1589; and I think it very probable that it was a gun taken from the wreck of a vessel belonging to the Spanish Armada, which was cast on the Scotch shores in the year 1584. Early in the following century, more serviceable, but still very rude guns, having the barrel attached to the stock by iron hoops, were introduced generally into Sutherland. These did not, however, entirely supplant the bow and arrow until after the middle of the seventeenth century. The spear was used at a more remote period against the boar and the wolf, and also in killing wounded deer.

The bow had one advantage over the gun, namely, that of being noiseless; so that, if the stalker were well concealed, he might repeat his shots without giving much alarm.

The sport afforded by the deer to the lovers of the chase with hound and horn (by which I mean hunting on scent, without the aid of rifles,) has always ranked high amongst the amusements of the upper ranks of civilised nations. In Germany, France, and England, up to a comparatively recent period, a pack of staghounds formed part of the establishment of every sovereign prince and wealthy noble; and this branch of the "Arte of Venerie" was reduced by rule and method almost to a science, and pursued in a stately and magnificent manner according to recognised principles, which are treated of at length in many works of the seventeenth and eighteenth centuries. But this is a field into which I have no intention of entering.

In this country, I believe, the stag is now hunted in his wild state only in Devonshire, and in the New Forest, and even there the animal is daily becoming scarcer. Staghunting was never probably practised in the same way in the Highlands of Scotland, the nature of the country offering great obstacles to this mode of pursuit on horseback.

It is mentioned in a letter, printed by the late Lord

---

* Sir Robert Gordon ascribes the name of *Glasnabhean* to John Beaton, the person who had charge of the gun, and used it in the skirmishes in company with Angus Baillie.

Graves, who hunted the wild-deer in Devonshire, that these animals, when they find themselves pursued by scent, generally run down wind; and the same thing has been asserted to me by others. This is an **extraordinary instance of** sagacity, as their natural instinct **leads them to the opposite** direction, it being a most difficult thing **for men** alone to drive them down **wind.**

In the following pages I confine myself to a description of the mode of killing deer now in use in the Highlands, which may be considered limited **to the** two methods of driving and stalking: the former of these offers more **room** for the companionship and friendly rivalry, which confers its main zest on sport of every description; but the latter, **if it has** the disadvantage of being pursued **in a more solitary guise, yet gives** so much scope **to skill and manœuvring, and exhibits the motions and the defensive** instincts **of the stag in such a beautiful manner, tried as he is under every variety of incident, that I have always considered it as creating a deeper and more absorbing interest.** Those sportsmen, therefore, **who agree with me, will not be** surprised at my dwelling **on it** with the natural preference of a fond lover of the spirit-stirring craft.

CHAPTER II.

Start from Blair **Castle.—Bruar Lodge.— A Comrade joins.— Ascent of Ben** Dairg.— Ptarmigans. — Forest Scenery. — Spirit-stirring **Interest.** — A **Hart** Discovered — Manœuvring.—Wading a Burn.—Getting a quiet Shot.—Dogs Slipped.—The Bay in a Mountain Cataract.—Dogs in Peril.—Death **and** Gralloching of the Deer.—Cruel Death of a Deerhound.—Origin and Antiseptic Property of Peat Bogs.—Ascent of Ben-y-venie.—A Herd Discovered.—Plan and Manœuvring.—Alarm, and Movements of the Deer.—An Injudicious Shot—A Successful One.—A Deerhound Slipped.— Bay.—Strange Adventure.—A Wild Huntsman.—Encounter with a Boar.—Loss of a Huge Salmon.—The Gown-cromb of Badenoch **and his** Story.

> " As he came down by Merriemas,
> And in the benty line,
> There has he espied a **deer lying**
> Aneath a bush of **ling."**
> *Minstrelsy of the Border.*

"It's just the skreek o' day, yer honour, and time ye war out o' bed."

"Rather farther on, I'm thinking, Peter; so take away the rifles, balls and all, get the men together, and make good speed over the moor: I see by the course of the clouds, which I have been watching from my bed, that the wind is in the right airt, and as the weather is warm, the deer will be far out on the tops of the hills; so we will leave Blair as soon as possible, breakfast at the lodge as usual, and go round the north of Ben Derig, that we may take all the ground, and not lose a chance. I expect to find a friend at Bruar Lodge, who was to come across the moors from the Badenoch country, and he slept there last night, if he did not miss the track, which you know is but a wild one. So order the pony to the door at four o'clock, and take care I do not pass you at Auld Heclan, as I did the day we killed the great deer; and I say, Peter, do not forget the whiskey."

"Na, na, I aye tak tent o' that. Did ye ever knaw me lave it ahint?"

"No, faith, to do you justice, your memory never fails you there; and you take care to refresh it pretty often. So off with you, my good fellow, and keep that laugh to enliven you on your way, for it is a long and dreary one."

It may be thought that Tortoise said this in a half intelligible, drowsy tone. Not a bit of it. An eager sportsman never sleeps or slumbers; or, if he does so by way of variety, he starts into life at once, and springs up from his bed as if the deer were actually before him: neither does he say, "Sandy, bring me the balls;" or, "Charlie, bring me my powder-flask," or my jacket, or my shoes, or anything else of the sort; for he has very methodically laid out all these things on his dressing-table over night with his own proper hands. To be dependent on others in these matters is exceedingly youthful: no, he trusts to no man's vigilance, but relies upon his own, and this is his system, not only in the camp, but in the field.

Mounted on his horse, Tortoise soon left the silent castle, and away he went, winding his rugged course through the forest of pines, some standing stately and dark in their verdure, others riven and blasted by the storm, their bare bones lying across his path, or driven crashing into the torrent below, where the waters of the Banavie come struggling through their rude barriers. The morn broke silvery and bright over the mountain top, just moving with her refreshing breath the light leaves of the birch and mountain ash, which were scattered about in nature's careless haste, hanging in graceful forms, and glittering with the falling dew-drop.

Now and then a roe sprang up from the bracken, in the secret glades of the wood, and vanished instantly with a bound among the gloom of the thickets, as the feet of the good Galloway clattered over the stones. To say that the rider "recked not of the scene so fair," were to do him injustice. No sudden gleam of light shot vividly across the moor,—no cataract leaped and dashed down the rocky chasms,—no wreaths of mists rose sluggishly to the mountain tops, with their light trains flickering behind, the effect of which did not excite his mind powerfully, and awaken it to the most pleasurable sensations.

"These are thy glorious works, Parent of good!"

So mused he thankful. At length, freed from the gloom of the pine woods, his pony dashed forward to the open moor, and the light spread broad around him; not a cloud is to be seen to indicate the course of the wind; a moment he checks his horse on the summit of the first hill, and scatters a few shreds of tow; away they sail to the north. Burn after burn is left behind him, but still he sees the long cart-track winding into the distance; and in the remote sky-line a few specks, which surely are his men, now near Bruar Lodge. At length the last hill is gained, and from its summit he descries the smoke issuing from the little chimneys; joyfully he sees it, and the complacent thoughts of breakfast come like balm over his soul.

There are some classes of men, alas, who know too well what hunger is: (would they were fewer)! Were I called upon off hand to name a few individuals particularly tortured by famine, why then I should say Franklin, Richardson, Ross, and the deer-stalker, who has ridden over moor and mountain, from Blair to Bruar Lodge, before sun-rise, conscientiously putting the deer-stalker decidedly first. Still let him beware of indulging his appetite too liberally. Something we shall say on this subject when we touch upon the necessary qualities and conduct of a deer-stalker. But, after all, what is the use of preaching up abstinence to a craving, ravenous mortal? Eat he inevitably will, and that to the last possible extremity, notwithstanding we tell him he may as well swallow coals of fire like Portia.

About eight reputed miles north of Blair Atholl, which distance would be numbered ten in a country of milestones, you descend into a glen, which is of a wild and desolate character. The heather being old, is rather of a brown than a purple colour; but there is some relief of green sward near the lodge, and more in various patches near the winding course of the Bruar. Huge, lofty, and in the district of Atholl, second only in magnitude to Ben-y-gloe, Ben **Dairg**, or the red mountain, stands dominant. At the right entrance of the pass, the little white and lonely dwelling, called Bruar Lodge, lies a mere speck beneath it. It consists of two small tenements facing each other, encompassed by a wall, so as to form a small court between them: one of these buildings serves for the master and the other for his servants; there is, besides, a lodging-place for the hill men, rather frail in structure, and a dog kennel of the same picturesque character. Close by stands a black stack of peats. Down winds the river Bruar through the glen, sometimes creeping silently through the mossy stones, and at others raving, maddening and bearing all before it, so that neither man nor beast may withstand its violence. Nearly in front of the little lodge is a wooden foot bridge, raised high above the water, so as to give it a free passage. When Tortoise flourished, this bridge, shot away by the floods, used to make an annual excursion of some miles towards the Garry, and was as regularly brought back again piecemeal, by a train of carts every summer. Like the boat-bridge on the Rhine, it might be termed a *pont volant*. Some distance up the glen, towards the east, a lofty cataract falls from the mountain side, whose waters find their way into the Bruar; and the head of the pass is obstructed by a chain of mountains, so that it forms a sort of *cul-de-sac*.

On these hills grouse are most abundant; and when they are not shrouded in mist, there cannot be a more delightful range for a sportsman. Tortoise, therefore, used to relax a little on them after the severer exercise of deer-stalking, when venison was plenty, and grouse scarce at the castle, or when the wind was unfavourable for the pursuit of the nobler game. By the favour of the lord of the forest, Bruar Lodge* was his occasional domicile. With all its apertures he loved it dearly; and it may be doubted whether any monarch ever entered a palace, or any lady a ball-room, with more absolute delight than he was wont to enter this lonely abode. What, though the winds would revel freely in it, and heave up the little carpet with an unceasing undulation, still the table cloth was tolerably tranquil, for the weight of the meal made it retain its station! What, though the parlour bell in the passage would ring incessantly during the night, even when the doors were closed, stimulated by the gentle violence of the wind; it was an Æolian harp to him! What, though a deluge of continuous rain, like the bursting of a waterspout, would sometimes plunge down, and darken the narrow glen, recalling the days of Deucalion and Pyrrha, still there was a to-morrow, and then the mist would climb the mountain tops, and the sun break forth anew in all its refulgence!

Heaven be praised for these transient checks, they add new vigour to our mind, and fresh zest to our sport.

But away with these reflections; for here comes my friend, safely arrived over the dubious tracts of the Badenoch mountains, fresh and eager for the sport.

"Well, Harry, I am delighted to see you arrived, and to welcome you to my cabin; how do you like our country; and how did you and your sheltie get across it?"

"Country! why it is a vast chaos of mountains, rocks, and torrents; I hit the track by a mere miracle,—you know that well enough. I am aware that the descendants of the

---

* The noble proprietor of Bruar Lodge would have spared no trouble or expense in making it as comfortable as possible for the writer of these pages, and this was repeatedly and kindly pressed upon him at Blair; but, as almost all his time during the shooting season was spent at the castle, he felt and expressed that every thing at the lodge was precisely as he could wish; and really, during a violent north wind and a raging tempest (the particular time alluded to), it did not come within the scope of a carpenter's or mason's craft to ward off the inroad of the elements.

Picts dwelt to the north, but without this previous instruction, I should be inclined to say, '*Nunc terminus Britanniæ patet ;—nulla jam ultra gens, nihil nisi fluctus et saxa ;*' so utterly desolate seems all around me. I dare say we shall see Galgacus in the course of the day. But pray let us go in; the breakfast is prepared, and has a most inviting appearance. Your men descried you on the last hill-top with their glasses, and all is ready. I never was more happy to see any one in my life; for besides other considerations, 'the air bites shrewdly;' and I am hungry past endurance. What a rascally hill that is at the head of the pass; my pony slid down it on his hocks, carrying forward with him a rattling mass of stones and rubbish, that now forms a talus, which, under ordinary circumstances, ought to have been the work of ages."

What was dispatched at breakfast we may not say; it becomes us not, when in our own cabin, to record how often empty plates were exchanged for full ones, or to say whether the pasty was still a venison pasty, or only a simple unpretending dish of earthenware; let those who have felt the mountain breeze, and all the freshness and salubrity it imparts, form their own conclusions; and we really can assure them that, all things considered, we are not voracious, that is, not particularly so,—on the contrary, we always feel inclined to inculcate the doctrine of abstinence; but then we cannot very decently do this to our own guests, so you must excuse us for the present.

"Now, Harry, are you ready?"

"In one moment,—just let me take another egg: and with your permission I will put this broiled grouse in my pocket, and a roll or two, and so forth. Now, then, for this wonderful work."

"Do you still hold your intention of taking no rifle?"

"For to-day, yes, most decidedly; I will keep cool, and see the nature of the thing first. That is my firm resolve."

"Well, I shall have three rifles with me, and you can use mine whenever you feel inclined to do so. I will explain the abstruse science to you, and all the meaning of our operations as clearly as I can, and I hope they will awaken your interest. The men are ready, and the dogs are in the leash, so let us sally forth. See, we must ascend this mountain; it is called Ben Dairg, which means the red hill; and, when we are near its summit, we shall be at the head of our cast."

"That will not take us long, I think, though it seems pretty steep; but the top is not far off."

"You cannot see the top from hence; but when we arrive at the point, which you mistake for it, which is a mere brae, the ascent is somewhat steeper, till you come to a naked point of rock, and sundry large uncomfortable stones."

"Well, thank heaven, there it ends at last."

"Wait a moment. Having reached this rock, a little cairn serves as a mark for our course, and guides us to the bare thin soil; and when we are at that spot, why, then, we shall see the top of the mountain. In fact, you must have seen it yourself yesterday, if it was clear, which I should doubt."

"I certainly did see a great mountain all the way before me, which blocked out the hills to the north, and grew bigger and bigger as I advanced, like a giant in a dream. A thick mist clung around its summit, and I pitied the poor eagles that were wheeling in the vapour. It made me dream of precipices and vultures all night long. You don't mean to say that we are to go there without a balloon. Why, Chimborazo is a mole-hill to it!"

"No, we shall not go to the very summit; but you are walking so stoutly, that I am sure you will not be the last of the party; and, to say truth, the mist that sits on the peak makes it look higher than it *really* is."

"Well, Davy, I see you have got Corrie and Tarff, and you are right, for that eager devil Ossian pulled so hard yesterday in the couples, that he must be quite unfit to go out to-day. It was worse for him than running ten chases; why, you could scarcely hold him."

"I dinna ken what sort of work it war to him, but I ken weel enough what it war to me, for he pulled me heels over head twice, in rinning down Ben-y-gloe, to turn the deer, him and anither, that's Oscar."

"To try to turn them, you mean, Davy, but they were over wilful, and gang'd their own way in spite of you."

The party were now breasting the mountain, and soon overcame the first ascent; when, turning to the left, they kept the northern side of Ben Dairg, and bore off towards the east, till they arrived under that huge mass of large gneiss and granite blocks which forms the summit of the mountain. The ground here was strewed over with the bones of calves (fawns), lambs, and moor-fowl, which had fallen a prey to the fox, wild cat, or eagle; and it was settled that traps should be set for the depredators.

"What! have you rabbits here? I thought I saw one run under the rocks."

"It must have been a white hare, which is nearly the colour of a rabbit in summer, and occasionally burrows like him. There are no rabbits here."

Lightfoot now suddenly seized the arm of his friend with an earnest look and panting heart, and making a signal for silence, pointed to a particular spot amidst the chaos of granite blocks. There was a sort of "air of success about him," that seemed to say he had made a capital hit; and, in truth, his excitement appeared to be excessive: judge, then, of his surprise and disappointment, when the only answer he got was,—"Ay, those are ptarmigans: you can have a a day at them when you have nothing better to do. They are not worth our notice at present,—guarda, e passa."

They now turned up the hill to the south-east, and proceeded till they came to an immense block of granite which stood upon the sky line of the hill; and then the gillies sat down on the heather;—he with the dogs in the leash, a little apart from the rest.

"Is this the forest? why, there is not a single tree or bush within ten miles of us."

"You are nearly right there, Harry; it is a forest only in the sense of the chase; wild as this immense tract is, however, every rock, corrie, cairn, and mountain is distinguished by some particular name, 'nullum sine nomine saxum;' and there are numerous sub-divisions which indicate every precise spot, so that the men appointed to bring home the dead deer, being thus told where they lie, never fail to find them."

"And now what do you think of this wild region? Do you not always feel as if you were wandering in a new world? Here, everything bears the original impress of nature, untouched by the hand of man since its creation. That vast moor spread out below you; this mass of huge mountains heaving up their crests around you; and those peaks in the distance, faint almost as the sky itself,—give the appearance of an extent boundless and sublime as the ocean. In such a place as this, the wild Indian might fancy himself on his own hunting grounds. Traverse all this desolate tract, and you shall find no dwelling, nor sheep, nor cow, nor horse, nor anything that can remind you of domestic life; you shall hear no sound but the rushing of the torrent, or the notes of the wild animals, the natural inhabitants; you shall see only the moor fowl and the plover flying before you from hillock to hillock, or the eagle soaring aloft with his eye to the sun, or his wings wet with mist.

"Nothing more shall you see, except the dun tenants of the waste, which we are in search of, and these I hope to fall in with long before we reach Blair. You have hitherto seen nothing but our tame deer, with their palmated branches, cooped up in ornamental parks; and such are picturesque enough; but when I show you a herd of these magnificent animals, with their pointed and wide-spreading antlers, ranging over this vast tract, free as the winds of heaven, I think you will agree with me that there does not exist a more splendid or beautiful animal; for whether he is picking his scant food on the mountain tops, or wandering in solitude through the birch groves, or cooling himself in the streams, he gives grace, character, and unity to every thing around him. How you feel I know not; but when I first trod these glorious hills, and breathed this pure air, I almost seemed to be entering upon a new state of existence. I felt an ardour and a sense of freedom that made me look back with something like contempt upon the tame and hedge-bound country of the South. Perhaps it is impolitic thus to raise your expectations as to the chase; and, indeed, it is impossible for me to describe the enthusiasm I felt when I first began my career. In the pursuit, the stag's motions are so noble, and his reasoning so acute, that, believe me, I had rather follow one hart from morning till night with the expectation of getting a shot (in which I might be probably defeated), than have the best day's sport with moor fowl that the hills could afford me. All your powers of body and mind are called into action, and if they are not properly exercised, the clever creature will inevitably defeat you: it is quite an affair of generalship; and if you have any thoughts of the army, I would advise you to scan all our motions, that you may gain a knowledge of ground and skirmishing.* You will find that almost every step we take has a meaning in it; we shall creep along crafty paths, between clefts and recesses, and make rapid and continuous runs according to the various motions of the quarry; so that when the deer are afoot, the interest and excitement will never flag for one single moment. See what a boundless field for action is here, and what a sense of power these rifles give you, which are fatal at such an immense distance! When you are in good training, and feel that you can command the deer, your bodily powers being equal to take every possible chance, the delight of this chase is excessive, as I trust you will ere long experience;—and here ends my eulogy."

"Well, I have listened to you with great interest, for I see your heart goes along with your words; and I burn with impatience to see a sport which every individual I have met on this side of the Tay seems to be perfectly wild about. Why, what a primitive country is this; are there any buffalos here?"

"Not exactly."

"Nor wolves?"

"Not at present; but sit you down quietly where you are, whilst we look for the deer: you may amuse yourself by eating the provender you put in your pocket at starting."

"No bad hint that; will you have a little? You won't; —oh, very well."

Tortoise and Peter Fraser now laid down their rifles on

---

* It is a fact that one of our most gallant and celebrated generals (why should I forbear to mention Lord Lynedoch?) declared that he got his knowledge of ground in this forest.

the heather, put their caps in their pocket, and crept forward on their hands and knees to a large granite block; then, cautiously peering over its summit, they began to examine the ground with their telescopes steadily poised upon it.—

"Well, Peter, I can see nothing but those eternal hinds on the Mealowr, and not a good hart amongst them: the ground is quite bare; so jump up, and let us get round the east of the Elrich, and see if there is anything in the corrie.—Maclaren, what are you glowring at?"

"Why, as sure as deid, I had a blink of a hart lying in the bog by the burn under the Mealowr. But my prospect is foul; he is lying beyond that great black place in the bog, joost in a line wi' thae hinds wha are on the scalp of the hill aboon."

"And a noble fellow he is, Maclaren; I can just see his horns and the point of his shoulders. It is a glorious chance; for, once in the burn, we can get within a hundred yards of him, and that is near enough in all conscience.— Here, Lightfoot, look at the fine fellow; pull off your cap, and rest the glass on the stone."

"Not the semblance of a deer can I see; but I'll take your word for it: I dare say he is there, since you say so. And now explain to me how you mean to get at him; communicate, my good fellow; for it seems, by all your caution, that even at this distance you dare not show a hair of your head."

"Creep back, then, behind the hill, whilst I mark the very spot in the burn which is opposite his lair.—Well, now I will tell you:

"We must go all round by the east behind yon hill, and then come up at the notch between yon two hills, which will bring us into the bog; we can then come forward up the burn under cover of its banks, and pass from thence into the bog again by a side wind, when we may take his broadside, and thus have at him. So let us make the best of our way. It would be quite easy to get at the hart, if it were not for the hinds on the top of the hill; but if we start them, and they go on belling, the hart will follow them, whether he sees us or not. Get our wind he cannot.

"Well, Lightfoot, you have come on capitally; and have

hitherto been able to walk like a man, with your face erect towards heaven. But now we are below the hill we must imitate quadrupeds, or even eels, for an hour or so. You have promised most faithfully to comply with my instructions; so, pray, walk and creep behind me, and carry yourself precisely as I do. Be like unto the dotterel, who, according to the worthy and veracious Camden, stretches out a wing when the fowler extends his arm, and advances his leg when the said fowler puts forth his corresponding limb. Above all, be as silent as the grave; and when you step upon stones, tread as lightly as a ghost. If your back aches insupportably, you may lie down and die; but do not raise yourself an inch to save your life, precious as it is. I assure you I am in earnest when I press all this on your attention, for it is absolutely necessary. Now let us put our caps in our pockets. Heaven bless me! do not raise up your hair with your fingers in that manner. I assure you, my good fellow, that just at present it would be much more becoming to be bald, or to wear your hair like King Otho.

"Maclaren, you will remain here, and watch the deer when I have fired. Sandy, follow you at a proper distance with the dogs; and come you along with us, Peter, and take the rifles. And now, my lads, be canny."

The party then advanced, sometimes on their hands and knees, through the deep seams of the bog, and again right up the middle of the burn, winding their cautious course according to the inequalities of the ground. Occasionally the seams led in an adverse direction, and then they were obliged to retrace their steps. This stealthy progress continued some time, till at length they came to some green sward, where the ground was not so favourable. Here was a great difficulty: it seemed barely possible to pass this small piece of ground without discovery. Fraser, aware of this, crept back, and explored the bog in a parallel direction, working his way like a mole, whilst the others remained prostrate. Returning all wet and bemired, his long serious face indicated a failure. This dangerous passage then was to be attempted, since there was no better means of approach. Tortoise, in low whispers, again entreated the strictest caution.

"Raise not a foot nor a hand; let not a hair of your head be seen; but, as you value sport, imitate my motions precisely: every thing depends upon this movement. This spot once passed successfully, we are safe from the hinds."

He then made a signal for Sandy to lie down with the dogs; and, placing himself flat on his stomach, began to worm his way close under the low ridge of the bog; imitated most correctly and beautifully by the rest of the party. The burn now came sheer up to intercept the passage, and formed a pool under the bank, running deep and drumly. The leader then turned his head round slightly, and passed his hand along the grass as a sign for Lightfoot to wreathe himself alongside of him.

"Now, my good fellow, no remedy. If you do not like a ducking, stay here; but for Heaven's sake, if you do remain, lie like a flounder till the shot is fired. Have no curiosity, I pray and beseech you; and speak, as I do, in a low whisper."

"Pshaw, I can follow wherever you go, and in the same position too."

"Bravo!—here goes then. But for Heaven's sake do not make a splash and noise in the water; but go in as quiet as a fish, and keep under the high bank, although it is deeper there. There is a great nicety in going in properly: that is the difficult point. I believe it must be head foremost; but we must take care to keep our heels down as we slide in, and not to wet the rifles.—Hist, Peter: here lay the rifles on the bank, and give them to me when I am in the burn."

Tortoise then worked half his body over the bank, and, stooping low, brought his hands up on a large granite stone in the burn, with his breast to the water, and drew the rest of his body after him as straight as he possibly could. He was then half immersed, and getting close under the bank, took the rifles. The rest followed admirably. In fact the water was not so deep as it appeared to be, being scarcely over the hips. They proceeded in this manner about twenty yards, when, the ground being more favourable, they were enabled to get on dry land.

"Do you think it will do?"

"Hush! hush!—he has not seen us yet; and yonder is my mark. The deer lies opposite it to the south: he is almost within gunshot even now."

A sign was given to Peter Fraser to come alongside, for they were arrived at the spot from which it was necessary to diverge into the moss. In breathless expectation they now turned to the eastward, and crept forward through the bog, to enable them to come in upon the flank of the hart, who was lying with his head up wind, and would thus present his broadside to the rifle when he started; whereas, if they had gone in straight behind him, his haunches would have been the only mark, and the shot would have been a a disgraceful one. Now came the anxious moment. Every thing hitherto had succeeded; much valuable time had been spent; they had gone forward in every possible position; their hands and knees buried in bogs, wreathing on their stomachs through the mire, or wading up the burns; and all this one brief moment might render futile, either by means of a single throb of the pulse in the act of firing, or a sudden rush of the deer, which would take him instantly out of sight. Tortoise raised his head slowly, slowly, but saw not the quarry. By degrees he looked an inch higher, when Peter plucked him suddenly by the arm, and pointed. The tops of his horns alone were to be seen above the hole in the bog; no more. Fraser looked anxious, for well he knew that the first spring would take the deer out of sight. A moment's pause, when the sportsman held up his rifle steadily above the position of the hart's body; then, making a slight ticking noise, up sprang the deer; as instantly the shot was fired, and crack went the ball right against his ribs, as he was making his rush. Sandy now ran forward with the dogs, but still as well concealed by the ground as he could manage.

"By heavens he is off, and you have missed him; and here am I, wet, tarred, and feathered, and all for nothing; and I suppose you call this sport. If you had killed that magnificent animal, I should have rejoiced in my plight; but to miss such a great beast as that!—Here, Peter, come and squeeze my clothes, and lay me out in the sun to dry. I never saw so base a shot."

"Hush, hush!—keep down. Why the deer's safe enough, Harry."

"By Jove, I think he is, for I see him going through the moss as comfortably as possible."

"We must louse a doug, sir, or he will gang forrat to the hill."

"Let go both of them; it will be a fine chance for the young dog; but get on a little first, and put him on the scent; the deer is so low in the bog that he cannot see him."

Fraser now went on with the hounds in the leash, sinking, and recóvering himself, and springing from the moss-hags, till the dogs caught sight of the hart, and they were slipped; but the fine fellow was soon out of the bog, and went over the top of the Mealowr. All went forward their best pace, plunging in and out of the black mire, till they came to the foot of the hill, and then with slackened pace went panting up its steep acclivity.

"Now, Sandy, run forward to the right, if you have a run in you, and get a view with the glass all down the burn of auld Heclan, and then come forwards towards Glen Deery if you do not see the bay there. Come along, Harry, the deer is shot through the body I tell you."

"Sangue di Diana! what makes him run so, then?"

"Hark! I thought I heard the bay under the hill.—No; 'twas the eagle; it may be he is watching for his prey. Hark again: do you hear them, Peter?"

"I didna hear naething but the plevar; sure he canna win farther forrat than auld Heclan; he was sair donnered at first, but he skelped it brawly afterwards: we shall see them at the downcome."

True enough they did; for when they passed over the hill to the south, the voice of the hounds broke full upon them, and they saw the magnificent creature standing on a narrow projecting ledge of rock within the cleft, and in the mid course of a mountain cataract; the upper fall plunged down behind him, and the water, coursing through his legs, dashed the spray and mist around him, and then at one leap went plumb down to the abyss below; the rocks closed in upon his flanks, and there he stood, bidding defiance in his own mountain hold.

Just at the edge of the precipice, and as it seemed on the very brink of eternity, the dogs were baying him furiously; one rush of the stag would have sent them down into the chasm; and in their fury they seemed wholly unconscious of their danger. All drew in their breath, and shuddered at the fatal chance that seemed momentarily about to take place.

Fortunately the stag (sensible perhaps of the extreme peril of his own situation) showed less fight than wounded deer are apt to do; still the suspense was painfully exciting, for the dogs were wholly at his mercy, and, as he menaced with his antlers, they retreated backwards within an inch of instant dissolution.

"For Heaven's sake, Lightfoot, stay quietly behind this knoll, whilst I creep in and finish him. A moment's delay may be fatal; I must make sure work, for if he is not killed outright, deer, dogs, and all, will inevitably roll over the horrid precipice together. Ah, my poor, gallant Derig!"

"May your hand be steady, and your aim true, for my nerves are on the rack, and yet I must own that it is the most magnificent sight I ever beheld; bayed by two furious animals, and with the death-shot in his fair body, the noble—the mighty hearted animal still bears up undaunted."

Tortoise listened not,—waited not for these remarks, but crept round cannily, cannily, towards the fatal spot, looking with extreme agitation at every motion of the dogs and deer; still he dared not hurry, though the moments were so precious.

Of the two dogs that were at bay, Derig was the most fierce and persevering; the younger one had seen but little sport, and waited at first upon the motions of the older, nay, the better soldier; but his spirit being at length thoroughly roused, he fought at last fearlessly and independently. Whenever the deer turned his antlers aside to gore Tarff, Derig seized the moment to fly at his throat, but the motions of the hart were so rapid that the hound was ever compelled to draw back, which retrograde motion brought him frequently to the very verge of the precipice, and it was probable, that as he always fronted the enemy, he knew not, or, in the heat of the combat, had forgotten the danger of his situation.

The stag at length, being maddened with these vexatious attacks, made a desperate stab at Derig, and, in avoiding it, the poor dog at length lost his footing,—his hind legs passed over the ledge of rock, and it now seemed impossible for him to recover himself.

His life hung in the balance, and the fatal scale appeared to preponderate. Still his fore legs bore upon the ledge, and he scraped and strove with them to the utmost; but, as he had little or no support behind, he was in the position of a drowning man, who attempts to get into a boat, and, being also, like him, exhausted, the chances were considerably against him. In struggling with his fore legs he appeared to advance a little, and then to slip back again, gasping painfully in the exertion; at length he probably found some slight bearing for the claws of his hind feet, and, to the inexpressible relief of every one, he once more recovered his footing, and sprang forward at the deer as rash and wrathful as ever.

Tortoise had at length gained the proper spot,—the rifle was then raised,—but when all hearts were beating high in sudden and nervous expectation of a happy issue, the dogs were unfortunately in such a position that a shot could not be fired from above without risk to one of them, and the danger was fearful as ever.

Three times was the aim thus taken and abandoned. At length an opening: the crack of the gun was heard faintly in the din of the waterfall;—the ball passed through the back of the deer's head, and down he dropped on the spot, without a struggle.

"Cadde, come corpo morto cade."

The dogs now rushed forward, and seized him by the throat;—so firm and savage was their grasp, that they were with difficulty choked off. The men came cautiously on the ledge of the rock, and began to take out the huge creature, two at his fore legs, and two at his hind quarters, and thus they lifted him out from the course of the torrent, and laid his dun length upon the moss.

"Ou, what a bowkit beast! Fende his haunches, and see sic a bonny head!"

"Ah, this is the best deer we have killed this year, Peter. I have not seem the like of him since the great monster I felled on the Elrich, when you put two charges of powder and two balls in my rifle; and the man who cuts up the deer so beautifully, at Blair, said he had a hole in his shoulder large enough to put his fist in."

"Will ye never forget that, then? But yer honour never held better, and sure oughten'd a big deer to hae a big load!"

"Admirably reasoned; I had forgotten that, Peter. Now, Lightfoot, what think you of deer-stalking?"

"Why, now we have got the deer, I must own it is most glorious sport; from the time we began imitating all the reptiles on the face of the earth, and bowing like the Persian, my heart was throbbing with excitement. It appeared as if all our craft and caution was to lead to some great end —an end not easily attained; which, you know, heightens the pleasure of success: and then the bay was sublime— positively awful! To be plain with you, however, I did not much relish gliding up the burn, trout fashion, not being gifted with fins. And now I am more than ever averse from Demaillet's theory, who conceived the globe to have been covered with water for many thousand years, and that, when the waters retreated, the inhabitants of the sea became terrestrial animals, and that man himself began his career as a fish."

"Well, we will have a good round of whiskey, and a health to the lord of the forest, who will smile when he sees this fine fellow. You got on most capitally."

"Why, yes, yes, pretty well over the moss-hags; but that confounded hill distressed me exceedingly;—that, and the grouse, mutton-chops, eggs, and rolls, venison pasty, etc., drew hard upon my wind, and I should not have been sorry to have gone all fours again. But I rallied capitally—did not I?"

"Rallied! why, I never saw you beat; and, to say the truth, these mountains are not so formidable as they appear to be. I have been more oppressed in walking over flats, mashing turnips with my feet, after those little birds called partridges, where the action of the muscles never changes,

DEER AT BAY IN A TORRENT.

than I have ever been on this varied ground, where the air is so refreshing and elastic."

"Well, now you shall see the last offices paid."

"Ah, that plunging of your man's long knife into his chest, which is followed by such a stream of blood, is a very kind one indeed."

The deer, after having been thus bled, was opened and gralloched.

"Eh, look to the white-puddins, sir, and see till the fat in his brisket and inside, and just pass your hand over his haunches. Lord, what a deer!"

Lightfoot felt his haunches as desired, and asserted that they were enormously fat, with as much confidence as Parson Trulliber would have done, though his conscience told him he knew nothing at all about the matter.

"Sandy, man, tak' the bag and white-puddins, and wash them weel at the fall, and bring the bag full of water, and we will rinse out his inside, and mak' clean work wi' him."

This being performed, they turned his head back on his shoulder, and covered it with peats, then shook over him a little gunpowder, and tied a black flag to his horns, to scare away the ravens. A few peats were heaped up in a conspicuous place at a little distance, as a mark to show where he was lying.

"A fair beginning; now for another round of whiskey, and then back to the head of our cast. As you complained of being blown in going up the Mealowr, I must tell you that there are some tracts of ground that are believed to be so much under the power of enchantment, that he who passes over any one of them would infallibly faint if he did not use something for the support of nature; it is therefore customary to carry a piece of bread in one's pocket to be eaten when one comes to what is called 'hungry ground.' You ate enough, to be sure, but it was at the wrong place."

"What a narrow escape Derig had! It reminds me of an event which happened in Sutherland in the Dirriemore Forest.

"A high-couraged dog was slipped after a deer among

the cliffs and crags on the eastern side of Klibreck. In the heat and recklessness of pursuit, he fell down a sloping but very steep precipice, and alighted on a narrow shelf formed by a projecting piece of rock—in fact, precisely in such a situation as my dogs were in, with the exception, that these could be approached on one side, whereas this poor creature could neither ascend the steep bank from which he tumbled down, nor find any practicable passage by which he could escape from his terrible position. The rocks opposed an insuperable obstruction from above, and the precipice menaced certain death below. There was no escape—no means of rescue; the spot could not be approached by man; and the poor animal, expecting that assistance from his master which it was impossible for him to afford, kept up a continual howling for succour during day and night. He continued to linger in this frightful prison for several days, and the sounds of his voice grew feebler and feebler, until they ended in a sharp kind of whistle, interrupted by vain efforts to break out into a bark. Every kind of project was considered, but no means could be devised to save him, for the ground was of such a nature, that no one could be lowered and pulled up by means of a rope. At length, the faint sounds ceased—his flesh was carried away by eagles, and his bones are still whitening on the rock.

"Now, Lightfoot, you are once more a free agent, and may get forward in the attitude most convenient to you; and pray talk as much as you please: '*minus via lœdat.*' We have no chance of seeing deer for some time, all this ground being disturbed."

"What! are we to go through that confounded peat-bog again?"

"Do not disparage it, for it abounds in grouse; and you see how useful its black channels proved in concealing us. I think its present state better for a sportsman than its original one; for, doubtless, it was formerly covered with trees, and the change has been brought about by their fall, and the stagnation of water caused by their trunks and branches obstructing the free drainage of the atmospheric waters, and thus giving rise, as you see, to a marsh: this,

Mr. Lyell has asserted of peat-mosses generally; and he mentions also particularly, 'that in Mar Forest, large trunks of Scotch fir, which had fallen from age and decay, were soon immured in peat, formed partly out of their perishing leaves and branches, and in part from the growth of other plants.' In the Forest of Atholl, we find everywhere in these bogs, roots of trees fixed to the subsoil, so that no doubt can exist of their having grown on the spot. My men dig some of them up annually, and they make excellent firewood, burning with great brilliancy, owing to the quantity of turpentine they contain. The eminent author I have quoted says also—'It is curious to reflect that considerable tracts have by these accidents been permanently sterilised, and that during a period when civilisation has been making a great progress, large areas of Europe have been rendered less capable of administering to the wants of man.'"

"I cannot quite assent to this latter remark of your eminent geologist, since I opine that venison and moor fowl, which the moss now nourishes, are incomparably better than oat cake and mutton, and that one of your fine, straight-limbed, sinewy Highlanders here are worth a thousand of such lazy fellows as Tityrus, and all that class of piping milksops:—aye, and Sir Walter Scott would have made them more poetical too, or, at least, more interesting. Hallo! by Jove I'm in for it."

"Heaven bless you! you should never put your foot in such a place as that, particularly when you are detracting from the Mantuan bard. Never mind, we will get you out presently. Here, Sandy, take you the right arm, whilst I lay hold of the other; now then—once—twice—thrice—and out you come, rather blacker to be sure, but quite as well as ever. Sandy, give Peter the dogs, and just scrape off the black dirt from Mr. Lightfoot with your deer knife, unless he wishes to enact the Moor of Venice."

(*Peter Fraser, touching his hat.*) "There's no such moor here awa', yer honour."

"These things will happen, but custom will make you better acquainted with such traps: let the ground look ever so bad, however, you may tread in perfect safety whenever you see stones lying about in it."

"Much obliged for your posthumous advice; but if I had been alone and had sunk in this bottomless bog, I should have been buried alive, and advertised for as missing."

"Something of that nature might probably have occurred; but I must tell you for your solace, in case of any future accident, that peat has wonderful antiseptic properties, and that you would have remained, though dead, in perfect preservation. Many instances are recorded of bodies so buried having been found fresh and unimpaired after a long lapse of years; and particularly the body of a woman was found six feet deep in the Isle of Anxholme in Lincolnshire: the antique sandals on her feet afforded evidence of her having been buried there for many ages; yet her hair, nails, and skin are described as having shown scarcely any marks of decay.* Thus you might have been exhumed after a few centuries, and put in a niche for the admiration of posterity, like the dried bodies at Monreale in Sicily, which are by no means alarmingly ugly, as I can testify."

"Highly alluring, certainly; I am glad, however, I was taken out for all that."

"Well, we shall now go along by the burn side, where the ground is firm, and then up that mountain which heaves its narrow back so high in the air. You have now seen what is termed a *quiet shot*; and I hope to show you sport of another description before we reach Blair, for all our best ground is to come. See, we are to go up this hill which leads to Cairn-Cherie; it will conduct us to the top of yonder grey summit, called Ben-y-venie, and there we shall have a fine command over all the deer that may chance to be within miles of it."

"Upon my word you try me hard, and, I believe, really wish to prove your peat's antiseptic qualities upon my frail body. The aërial perspective of that mountain's crest is

---

* Lyell's Geology.

exceedingly alarming; your soil is culpably ambitious and aspiring—

———— Superas evadere ad Anras,
Hoc opus, hic labor est.

I thought myself as good as any of you at first, but that struggle up the Meal-ower (I think you call it) undeceived me. A hundred yards of such a steep is, as Falstaff says, 'three score and ten miles to me.' But by Jove I'll have a pull for it; *andiamo dunque, andiamo pure*, and now beat me again, if you can."

The party proceeded obliquely up the hill eastwards, the files covering each other, and all masking themselves as much as possible behind knolls and blocks of gneiss or granite, under cover of which they repeatedly examined the country with their glasses. Had the fate of a whole army been dependent upon discovering and circumventing an ambuscade, no better tact or caution could have been observed. And now they had just gained such an ascendancy of the mountain, as would enable them to examine Glen Mark and the hill side beyond it, called Sroin-a-chro. This was an anxious time for the ground was so precipitous, or, in other words, so favourable for the sport, and Tortoise was so intimately acquainted with it, that good success might be expected if there was no lack of deer. The little party took care to keep below the sky lines; and all lay down in the heather except Tortoise and Fraser, who crept forward on their hands and knees without their caps, and then extended themselves on the ground, resting their glasses on the little eminence in front of them: these they moved slowly and steadily to all the favourite spots.

"Nae thing can I see forebye a few hinds on the Craggan-breach. Surely the glen can no want for harts?"

"Heaven forbid, Peter; but I fear it does, unless they are lying further on. There is a great deal of ground which we cannot see from hence, you know."

Fraser now looked intently for a long while at the same spot, and would pay no heed to any thing that was said to him. But when at length he turned back his head, there was such a relaxing smile on his face, as made it perfectly beautiful to a sportsman. These, indeed, were Peter's handsome moments,—illuminations that shot across his countenance like the sun-gleam on the moor.

"Now, where are they, Peter? for I see you have found them at last. Your eyes are ever the best."

"Creep back,—low, low. They are lying in yon corrie, rather high up. Hey, what fine harts! Ane, twa, three, four; there are eight a'-the-gither; twa of them are royal, and twa mair there are wi' wide heads and few branches, and these, I ken, are the fattest and bonniest of the lot: haud weel to them, Sir, if you have a chance."

"Never fear. Ah, now I see them. You said nothing about the hinds, whereof there are several; and one nasty, lop-eared imp there is, some way to the south, before the rest; and if we are foiled, as I fear we shall be, this beast will do it, for she was born for mischief."

"Hist, hist, Maclaren, come you here. Take the glass and examine the deer well, and most particularly that sentinel to the south, for she is the beast you must dress to when you start the deer. Take care and be well forward when you show above her, but so that the harts in the rear of the parcel do not get your wind. But it is useless to give you any instructions, for you know what to do as well as I can tell you; only take care they do not go tailing down the glen, and break off over Aukmark-moor. The wind you ken is full south, and a difficult job it will be to make them cross."

Maclaren looked long and intently at the deer, and not only ascertained their exact position, but examined all the rest of the ground, to see if there were any other deer that were likely to join them. He then sat down with a thoughtful countenance, every now and then plucking little pieces of grass, biting them, and flinging them away, like one in perplexity.

"I'm thinking it'll be no that aisy to get them ower Ben-y-venie; but I shall try to pit them intill your ground at ony gait. The beast will be unco kittle to dale with. Ye'll be patient, Sir, and gie me time."

"If they do not come it will not be for lack of skill, or good will on your part, Mac, for a more clever or willing man never trod the hills,—in sight and out of it, alike to be depended upon."

"Now, Davy, a word with you. What is that sticking out in the right pocket of your jacket?"

"That's joost the whiskey."

"And what is that great lump in your left pocket?"

"That's in my left! Why, then, that's joost the ither whiskey."

"But you seem to have something pretty considerable in the right pocket of your trowsers; what may that be, Davy?"

"That's the wee bit pewter whiskey flask, yer honour."

"Then that protuberance opposite, on the left?"

"Why sure is'nt that the ither pewter flask?"

"Well, Davy, thou art most judiciously balanced, and thy providence is much to be commended; just take out one of the large bottles, and let us see what it is like. Now for the pewter cups, and fill round to every one, that they may drink good success to our manœuvres. You are a perfect walking cellar, Davy; how many bins of whiskey you have about you I cannot precisely say; but we will have compassion on you, for at any rate you are heavily laden. Just give one of the flasks to Peter Maclaren;— nay, give it man, and leave a black bottle with Sandy; and now to your posts. Sandy, set you off for Ben-y-chait."

"Upon my word, Mr. General Tortoise, you are a very mysterious person; I have listened very attentively to all you have said, and silent I have been, as not presuming to interrupt the jargon of so consummate a general. As for the deer, I do not see them, though I have been looking through the glass this long while; but it seems you are going to put some manœuvre in practice, and I will thank you to tell me what your exquisite plan may be. You don't mean to say that you can get near deer in such an open country as this?"

"That is as it may be: we shall have to wait here about forty minutes, when I will disclose and illustrate; but I must first start Peter Maclaren. Now take your whiskey, and away with you, Peter."

Away went the clean-limbed hill-man down the mountain, skipping over the hillocks, diving, vanishing, and reappearing with a bound upon the moss-hags, like a stone hurled downwards in pure pastime. Arrived in the glen, he kept twisting and lurching in the darkest coloured ground, and, by making a circuit, managed to cross the stream out of sight of the game. Here we will leave him for the present, full of the importance of his embassy, and sensible that all his movements would be seen and canvassed.

While the sportsmen were lying down in the heather awaiting the event of Maclaren's mission, Tortoise pointed out the various features and nature of the wild tract of country that lay around them.

"We are now," says he, "on Ben-y-venie, which means the middle hill, or if you delight more in its other appellation, on Beinn-a-Wheadhounedh. That bulky, round headed mountain to the right is Ben-y-chait, from which we are separated by Glen Dirie. The mountain tract to the left consists of Craggan-breach, Sroin-a-chro, and Cairn-marnach. And this deep glen to the east is Glen Mark. You see by the indistinctness of the objects, how deep it lies beneath us; the river that runs through it in beautiful curves, as if loth to leave the solitary pass, is called the Mark: listen attentively, and you will hear a faint, hollow noise coming up the glen from afar; this is the sound of its waters falling into the Tilt. Some few miles away to the south, it forces its passage through a gloomy channel between the mountain crags, then dives through groves of birchwood; after which begins its ceaseless toil,—it rushes headlong into the Tilt,—for ever doomed to struggle with still more turbulent waters.

"Beyond these glens and mountains, many a mile and many a hill top lie between us and the end of our cast, and the whole is terminated by large pine woods.

"So much for our ground. You will soon see what we are attempting to do with those deer. In sportsman's language we have the command of this mountain, as well as of the glens and hill-sides on each hand of us, or at least we shall have it, when the men are arrived at their posts; for one of them will be on Ben-y-chait, on our right, and the other on Sroin-a-chro, on our left: we shall remain on this hill in the centre, and they will endeavour to put the deer on our hill. This, it is evident from the wild and open nature of the country, cannot be done by actual driving, but depends entirely upon skilful manœuvring, which I do not endeavour to explain at present, because you are about to see it put in execution.

"Do you see Maclaren, Peter Fraser?"

"He has louped the burn, and is in the moss forenent the crags."

"Now, as I was saying, Harry, I have not much hope that we shall get at these harts, but I make it a rule to try every possible chance. If we get them on our ground once, it shall go hard but we will keep them there the whole of the day. I think you will find this stalking in double quick time far more beautiful and exciting than the getting a quiet shot."

"Is Maclaren behind the hill, Peter?"

"No, no, he canna be that far as yet. You ken that yoursel'."

"That getting a quiet shot, Harry, has its charms, I must confess: the threading of the winding passages through bogs, up watercourses, and secret places in every possible attitude, except that adapted to the nature of a two-legged animal, is certainly picturesque and exciting. But then it is a sort of assassination; and you never get the intellect of the animal to bear against you, or see his motions, but steal upon him like a thief."

"For heaven's sake, my good friend, do not prose any longer, but tell me at once how the deuce we who are sitting here have any chance of getting a shot at those deer which are fifty miles beyond us. I long to be in action."

"Adagio, adagio, you shall see. Do not be impatient, my good fellow; I will not be chary of instruction when time shall serve.—Why, Peter, what the deuce is Maclaren about; will he never get behind the hill: are we to be kept here all day?"

"Why sure ye'll no be expectin' he'll be there the noo: he canna win that far in twanty minutes."

"Well, well; the time seemed longer."

"So, as I was saying, Lightfoot, you must not in this

case be impatient, but rather imitate the discreet Fabius. He would have been a capital hand at a quiet shot."

"Aye, and a capital proser too. But will you not give me leave to imitate you, my incomprehensible master, who have been fidgetting about, looking at your watch, taking up your rifles, and putting them down again a hundred times, and are as restless as a hyæna in a cage? A pretty sort of Fabius you are yourself."

"No, no, never mind me; it's only a way I have: or perhaps I consider patience as King Charles did morality: he loved it, he said, though he did not practise it. But I would advise you to ——. By the powers! I see him now; he is sitting down above the deer, and examining them with his glass. What a capital fellow; he has not been more than half an hour. Now he is looking at us for a signal: open your waistcoat, and show your shirt, Peter. —He sees it: now he is going forward behind the hill, and will soon start them."

"Lightfoot, come you here, and observe the beautiful motions of these animals, which to me are as entertaining as any part of the sport; but should the deer come near us, pray be mute as a fish, and as quiet as the most magnanimous mouse; keeping your hair smoothed down like unto those fair nymphs at Portsmouth, beloved of the sailors, who comb it straight in front, and cut it to the pattern of a bowl-dish."

"Now, take my glass—one of Dolland's best, it is—stay, I will direct it to the proper spot: look intently—keep the glass as steady as possible—and when the deer are in motion, and group together, you will be sure to distinguish them, though they are not so easily seen at present."

"Now, indeed, I do actually see them; what beautiful creatures! They are all standing up, and gazing at the summit of the hill. How stately the stags look with their jutting necks and towering antlers. Are you sure they are not elks? Gad, I think they are. How they are moving forward to the hind in advance, which you seem to have such an antipathy to. What in the world makes them shift their quarters?"

"Why, Maclaren is nearly opposite to them, but at a

great distance above, behind the swell of the hill, and doubtless has just shown them the top of his bonnet over the sky-line; but they are all going wrong, and do not seem inclined to accommodate us."

"They are not much alarmed, I think, for now they are standing still, and the hind has walked back a few paces, and is gazing up the hill again; the others seem to watch her motions, and to be guided by her judgment; whilst the harts appear to give themselves very little trouble about the matter."

"No, the lazy rascals! but we may rouse yet. Yes, they are alarmed, or, more properly speaking, suspicious. They have that sort of discretion which makes them run away in cases of danger; but you can never frighten them out of their wits with so small a force as ours. They are deliberately trying to make out what is going on before they decide upon the direction of their retreat, and are too proud to fly without evident cause. But just keep your eye upon them; Maclaren will not let them off thus; he will make a push for it at any rate."

And so it seems he did; for in a few minutes they turned aside, and came a little way down the hill, gazing in a fresh direction more towards the south.

"By Jove, they are turning!—capital!—well done, Maclaren!"

"Why how the deuce now did he manage that; and what has made them alter their course? Why, your men are almost as clever as the deer: upon my life this is very entertaining, especially now the herd are coming towards us; I feel my heart rioting and beating against the heather."

"Doubtless, when he saw the deer going southwards, he slipped back cannily behind the hill, ran like an antelope, and then came in again over the sky-line, and showed himself partially more in front of them. Faith, I see him now with my glass sitting very composedly on that crag that hangs over the glen; his legs seem to be dangling in mid air. That is right, Maclaren; let well alone. The deer cannot see you, I know, my man, though we can. One point at least is now gained; for I am happy to tell you they will never resume their first direction, for the slight dubious glimpse they had of the hill-man's bonnet makes them suspect an ambush in that quarter; but when they descend into this glen, which, as you see, lies some three thousand feet below us, they may go straight forward to the south, which will be equally bad, avoid our hill entirely, and extricate themselves from the Caudine Forks without a shot. But I hope Maclaren may match them yet."

"You will think this is slow work, and so indeed it is just at present; but if things go favourably, take my word for it, you will have no reason to complain on that score. We shall try your wind again, my good fellow, I promise you. But at any rate it is no little matter to see the graceful motions of the deer, and mark their intelligence and sagacity. See, now they stop, and examine all the glen before they venture rashly into it; they scan every part of the ground, and gaze so intently that no object can escape them that lies within the limit of their vision.

"I may as well tell you, that if the hill-man had come down right upon them in the first or second instance, and endeavoured to drive them as one drives sheep, they would immediately have raced away straight south, right up the wind, and have soon been out of our cast. When they see their enemy, they easily discover his drift, and take pretty good care to defeat it. See how carefully they march, like a retreating army, with their front and rear guard."

"Beautiful! and with such measured steps: so stately, winding down that horrid rocky precipice, which I should have thought impassable by living beast.—What are our firmest resolves? I shall take one of the rifles, if they come near enough, notwithstanding my previous determination, for this day I mean to immortalise myself."

"I am rejoiced to hear you say so: and now we must crawl farther forward, for the deer are fast sinking below out of our sight; already they are at the bottom of the glen, on the banks of the Mark; and now, Peter, after all this trouble, I fear our chance is gone, for they are all going straight down the glen, and will not cross to us."

Here Peter pressed the master's arm, and pointed.

"Did you no see yon parcel of hinds there towards the

shank of our hill? they canna chuse but join them, and they will come; but it will be low doon."

And now the skilful missionary, who had a clear and commanding view of all these things, began to set to work in a more determined manner; he pressed forward rapidly, still out of sight of both parcels of deer; till at length, when he came sufficiently forward, he dashed down the hill in full view, shouting, hallooing, and hurling stones down the mountain with all his might,—going to and fro as the deer shifted,—slipping, clambering, and tumbling, in such perilous places as would have endangered the life of a mountain goat. Greatly to be feared he was, as Polyphemus, when he hurled the rock at the Sicilian lovers; but not Maclaren, or Polypheme himself could have put these reasoning animals into any state of confusion; for, being too distant from the tumult to be under any apprehension of immediate danger, they continued to be perfectly deliberate in all their movements: it was like calm dignity opposed to passion.

The hinds last mentioned, which were opposite them, on Ben-y-venie, collected and wheeled about, much admiring what all these strange noises might portend. Now had the decisive moment arrived when the thing must terminate either one way or the other.

But let us see what the rifle-men are about. When they saw the hill-man storming, and heard the stones coursing each other down the crags, they were aware that no time was to be lost. Tortoise pressed his friend's arm:—

"Now, then, or never!—creep back quickly, and prepare for action; for, by Herne the hunter, they are coming; low, low, for heaven's sake! We must get on to that large stone, and they will all come into our very mouths. Now, then, forward! take this rifle, and hold well at the best antlers when time shall serve; be steady, and fire well forward, taking care not to drop the gun when you pull the trigger. By Jove! I see the points of their horns. Run low,—low, for heaven's sake! this is not our time. Hark, I hear them in the crags."

The faint clatter of their hoofs was indeed heard by all, as they were picking their way obliquely along the rocky ridge; and the stones that they put in motion coursed each other down the steep, and gave forth a sound, which, becoming fainter and fainter, died gradually away, as they rolled into the depth below.

But how uncertain are all the chances of the chase! How Fortune loves to baffle us! and how wise the Romans were to worship her as a deity, and erect a temple to her honour!

The goddess sulked unpropitious, and her frowns were met by Tortoise with the following eloquent exclamations, uttered, as was meet, *sotto voce*—

"Death and destruction! they are turning away. Oh, what a fine chance lost; they were coming up so beautifully! Confound ye all, ye regular set of misbegotten imps, don't you know your own minds? But you shall have a run for it yet.—*Guai a voi, anime prave!*

"Come along your best pace, Harry, for the hinds are started, and our parcel is racing up to them; keep you above me, which will save you ground; and, Peter, do you stalk the deer, and I will stalk you, which will give me a pull also. We will make a push for it yet."

In pursuance of this arrangement, Fraser peered down at the deer's horns, over the ribs of the hill side, ducking, skipping, and running, so as to keep out of their sight, and nearly along side of them,—the riflemen above keeping parallel to him, and dressing according to his motions. The deer, however, were steady to their tactics, for they were resolved not to come over the steep part of the hill, where, by losing the wind, they might come unawares on an enemy; thus they were rapidly advancing towards the foot of the hill, where the slope was so open and gradual that they could see a long way in advance, and consequently could not be suddenly surprised.

When Tortoise saw how unfavourably things continued to go, he persisted no longer in the same direction, which would only have given the deer a fresh start, and hurried them on to an impracticable distance, without any possible chance of his coming within shot of them.

Thus, whilst there was yet time, he turned suddenly to the right, and went rapidly over the hill in a new direction;

for as the herd had never seen him or any of his party, he judged they would remain for some time at least on the round swell of the hill below, which they were now approaching.

This continued exertion was a severe draught upon the vigour of the party; deplored by all, but by none more deeply than by the newly initiated sportsman; in fact, he was wholly unequal to it,—his limbs faltered, his knees trembled, and his breath came short and loud, till, quite exhausted, he lay down on the moor a solitary and forsaken man, while his inhuman companions persisted in their course. His spirit, however, was unbroken; for as soon as his wind was a little recruited, he got up and followed in the line.

And now Tortoise and Peter Fraser had reached the crags on the opposite side of the hill, towards the west. Here was an absolute precipice, and large angular stones were lying down it, with their edges uppermost. Happy was the foot that did not slide down upon their sharp ridges, and charmed was the leg that was not either cut or broken by them. The two practised hill-men, nothing dizzy, picked up their legs like cats, and went down pretty fast; having once begun the descent, indeed, it was not very easy to stop, so headlong was the steep.

And here I am sorry to be obliged to relate a circumstance that, for the sake of their credit, I would gladly have concealed, namely, that, from the time of their first rapid start, they never once took care of their companion, and, indeed, had as completely forgotten him as if he had never been of their party; so absorbed had they been in stalking, and so absolutely necessary was it for them to act precisely as they had done, or to throw away a capital chance.

The struggle now was to get under the hill, on the side opposite to that part which the deer were crossing, so as to arrive there in time to take them as they passed down over the boll of it, still preserving the wind. Arrived at length at this desired spot, breathless, flushed, and covered with perspiration, they crept forward and wormed themselves through the heather, till, from behind a small knoll, they saw the deer feeding forward very leisurely, but still restless, and with their sentinels looking back towards the east. And now the heat of the manœuvre being ended, they began once more to think of Lightfoot; and Tortoise, putting his mouth close to Peter Fraser's ear, as he lay on the ground beside him, desired him in a low whisper to beckon him alongside of them. "Here is a glorious chance!" said he, "and I would not have him lose it on any account."

"And it's mair the pity he's no here to tak the chance; but I have been speering aboot, and canna light on him. Sure as deith, then, but I see him the noo! eh, that's him, high up in the crags. Lord, Lord! what shall we do? it is an unco' fashious place for a stranger: he canna win forrat by himsel at ony gait."

"We should have considered that before, Peter; but creep back, and send Davy after him, with a caution how to bring him into the ground properly. The dogs will be back in time; and I trust he may yet join us before the deer cross. Speed, Davy, speed!"

Away went Davy over moss and crag, and up the steep, waving his bonnet to the vexed sportsman; but there was no charm in Davy's signs sufficiently powerful to induce Lightfoot to alter that method of descent which he himself judged most conducive to the preservation of his existence. In vain did the herald keep sawing the air with his bonnet, still advancing to the rescue. Our hero found his head swimming, and very wisely gave up the upright position, and made his way on his hands and knees, as best befitted his unhappy condition. At length the messenger reached and assisted him, and the crags once passed, both came forward rapidly.

Fraser, who had been peeping from time to time through a bunch of heather, now pressed Tortoise's arm and whispered, "Be ready,—they are coming!" Both were lying flat on the heather, with the rifles on the ground, on one of which Tortoise had his hand; but, as yet, he did not raise it. They lay still as death till some hinds passed within an easy shot; next came a four-year-old hart, which was suffered to pass also: the better harts were following in the same direction, and the points of their horns were just coming in sight, when lo! Lightfoot, who had that moment come into the ground, fired at the small hart which was galloping away gaily, and gaily did he still continue to gallop. This injudicious shot (which of course turned the other deer) struck woe and dismay into the soul of Tortoise; up he sprang, and dashed forward, but it was only to see an antler or two vanishing out of sight under the swell of the ground; still he went on as fleetly as ever he ran in his life, cutting off to the point where he expected the deer would reappear in crossing the bottom. There he arrived just in time to get a long shot at the last deer that was passing. He stopt short as an Arab's courser, and, standing at once firm and collected, took a deliberate aim at him. The crack of the ball could not be mistaken; it was that particular smack which it makes, distinct from any other, when a deer is stricken.

Davy came forward with the dogs at the well-known sound, followed by Lightfoot; the whole party then lay quietly down in the heather, Peter Fraser being enjoined to examine the herd as they passed up the opposite heights, and keep his eye on the wounded hart. This is always the surest way of recovering him, for if you press him, and he is not hit deadly, he will get forward in the middle of the herd, whilst his wound is fresh, and run with the other deer, in such a manner as will most probably occasion you to lose him; but, on the contrary, when he is not urged forward, and sees no one in pursuit of him, his wounded part stiffens, and he seeks ease by slackening his pace, or, if badly wounded, by falling out altogether from the rest of the herd; and if he is not badly wounded, you must lose him at any rate,—at least you will have no better chance with him than with his companions.

"Now tell me, my way-worn and much injured friend, what made you shoot at that little deer?"

"A little deer! a little deer! *haud credo*—I thought he was an enormous monster!"

"I must reply as Master Dull, the constable, did to the erudite Holofernes,—''Twas not a *haud credo*, 'twas a pricket.' Extremely juvenile he is, I promise you; but you will soon distinguish better. It would have been a dead loss to the forest to have slain him, for his flesh now is worthless; whereas, in two years more, he will be fine venison. But I would have borne all the blame at the castle, in requital for your good temper in not scolding me for leaving you on the crags of Ben-y-venie. But hinds and harts wait for no man; and, moreover, I should have given up a fair chance had I waited, without conferring any benefit upon you."

"Aye, food for eagles I might have been. All fair, all fair;—I undertook to follow you, and could not, that's all; and, to do you justice, you never looked behind. 'You have a straight back, Hal, and care not who sees it.' I am convinced that you have cloven feet, like Pan, or that fellow with a worse name (whom, out of deference to you I forbear to mention), or you never could have galloped down that fearful precipice like a chamois. It made me giddy at once; my head reeled, and I was a lost man—an absolute nonentity, wounded and heart-broken."

"And heartily glad am I that you are found again; without bruises you intimate, I may not say, but without broken bones at least I may, at any rate. But console yourself; you are not to blame, but rather your half-boots. Get the proper material in future,—thick shoes with nails, or Scotch brogues—

> 'The hardy brogue, a' sewed wi' whang,
> With London shoes can bide the bang,
> O'er moss and muir with them to gang.' *

'By the foot of Pharaoh,' as Captain Bobadil says, but this must be amended.

"Peter, do you see the wounded deer amongst the lot which are foremost?"

"Na, na, he's no there; he'll be coming up ahint."

"Give me the glass; I see him plainly enough: he is shot through the body, rather far behind, and cannot go far. Now one of the deer is licking his wound—now he begins to falter—now he turns aside, and sends a wistful look after his companions, who are fast leaving him, happy and free as the air we breathe. He is making another effort to regain them—poor fellow! it may not be; you shall never

---

* Galloway's Poems.

join them more. **Never again** shall you roam with them over the grey mountains—never more brave the storm together—sun your red flanks in the corrie—or go panting down to your wonted streams;—'brief has been your dwelling on the moor!'

"And now I am resolutely determined **never to fire at a deer** again—no, never whilst I live. It is a barbarous and inhuman practice; the act of **a savage, and** ought to be punished by branding, hanging, or at least by transportation for life. There—*(flings down his rifle)*—lie there, thou villain! '*hic cestus artemque repono.*'"

"They're a' ganging right, yer Honour, and we shall have them again beyond **Cairn Dairg Moor**."

"By Jupiter! so we shall, Peter. Here, **give me my rifle,** most humane of men, and I will aye make a clean **shot in future**."

"And I have seen you mak' clean shots half through the season; but the wee bit ball will whiles tak' his ain course;—naething mair wilfu'."

"Now then, Peter, take Percy, and get the wounded hart to bay: a fine fellow he is. I need not caution you **to** pass the scent of the herd before you lay him on. There is no hurry. In the mean time I will load my rifle myself, and then, Peter (you ken what I mean), we shall have no more broken ramrods."

"Did I brak a ramrod **since last Tuesday?**"

"Indeed you did not, my good fellow; **you** only rendered a powder-horn **unfit for service;** but I would rather have my ramrods broken daily, in the excitement and hurry of the moment **by a dear** lover **of the sport** like you, than have my rifles **loaded** carefully, slowly, and **mechanically**, by a tame and **lukewarm sportsman**. Here, take a glass of whiskey, Peter.

"Now, Lightfoot, we **will** wait here till we see the dog laid on. I am vain, you know, of **my** hounds, and Percy is **one of my best**. You see what **a pace Peter is** going with my favourite in the leash pulling him onward all the way;—**now they are** dashing through the stream—now he breasts the hill, and has passed the track of the herd, and is trying to find the slot of the wounded deer—he has it!—Percy scents

7

it too, and pulls down the leash, straining his nose to the ground;—do look at the eager fellow!

"He is slipped, and has overrun the scent: see what a cast he makes, with all the dash of a foxhound united to the speed of a greyhound:—beautiful!—there—he has it, and the deer is before him, going down towards the Tilt: come along, then; and follow you, Davy, with the other dog."

Off ran the sportsmen to the river Mark at their best enduring speed, and so on to the Tilt, where they expected at once to find the bay, but they were wofully mistaken. After having followed the wild romantic course of that impetuous torrent for some time, they overtook Peter Fraser, who seemed as much at a loss as themselves; still they kept running on, and at length came upon the track through a birch grove. Here and there they found the grey stones dyed with drops of blood: now, all were sure they heard the baying of the hound; but, although they kept advancing with their utmost speed over rock and ridge, through burn and cataract, it died away and was lost: again it was renewed; and the sound ceased as before. This was very strange! what should make a stag so badly wounded break his bay in such a manner? But Percy would never leave him, come what might. Once more, in rounding a point, they heard the bay distinctly, and not far distant. They gained upon it, and soon the fatal truth broke upon them, filled them with astonishment. Could it have been believed that, amongst the lonely woods of Glen Tilt, reserved alone for ducal sports—sacred as the harem: where neither stranger nor traveller were permitted to put a foot unbidden—in a country where the chase, and its customs, and its laws, were so well recognised and understood—could it have been believed, I say, that a mortal could be found so rash as to constitute himself the lord of the chase, setting aside the laws of the Medes and Persians? Yet there figured such a monument of audacity. He seemed to be a young man; certainly he had all the vigour and activity of youth. He shouted with all his might, rushed into the water, assailed the deer with stones, and tried to get in upon him and fell him with a sort of bludgeon which he brandished. A kilted Highlander was running towards

him, and, as it seemed, endeavouring to call him off; then came forth a general shout of invective from all the party as they ran forward. High above the rest rose the guttural sounds of the iracund forester.

In the midst of this tumult the hart broke bay, laboured out from the Tilt, and went heavily along through the birchen grove, being evidently much exhausted. Percy followed close upon his traces; then came the wild huntsman with whoop and hallo, dashing over knoll and rock, through bog and through burn, till he fairly vanished from the view.

"Contremuit nemus, et silvæ intonuere profundæ."

"The man's dementit. But sure it's na man, ava'; it's joost the kelpie; him that left the print of his fut on the muckle stane up bye forenent the Tilt, where he grapt the deer; and the deer's fut is there, too,—ye'll ha' seen it yoursel' sir."*

Toiling and jaded, the sportsmen followed as best they might, replete with wrath, and venting threats of vengeance from time to time as their breath permitted; but not one inch could they gain on the fleet-footed stranger. They came up with the Highlander, however, and made him go on with them as a prisoner. A word or two passed between him and the hill-man, who, it seems, knew him.

Percy's deep tongue again echoed through the pass, and it was hoped that the bay would last long enough to allow them to come up; if it did not, they had no expectation of outrunning a being whom some of the party took to be supernatural.

At length the stag was quite exhausted, and stood again at bay in the midst of the rushing waters. Always foremost, superior to every obstacle, and flaming with ardour, in plunged the reckless sportsman, intent, as it seemed, on close combat. Already was he making his approaches with uplifted club, when Tortoise, who had gained upon him during the bay, raised his rifle from a distance,—the ball whizzed close by the assailant, and down floated the mighty hart, a lifeless thing!

---

* These impressions actually exist at present, quite perfect, in the place alluded to.

The stranger splashed after him, rushed at him, and was the first to grip him and drag him towards the shore, till the hill-men came up and took the affair into their own hands.

When protracted torments, however acute, terminate in complete success, it is astonishing how suddenly all preconceived anger ends with them. *Considunt venti fugiuntque nubes.* Thus it was with Tortoise; and when he saw the open, happy countenance of the English stranger, who accosted him as if he had performed the most serviceable feat in the world, he could not forbear laughing outright.

"Fine sport, sir," said the wild huntsman;—"glorious sport!—but you finished it a little too soon; I would you had let me come at him again,—I would fain have plucked the laurel."

"I believe, sir, we are indebted to you for having protracted the good sport so long; for owing to your very valorous exertion we have pursued that noble fellow some miles farther than we had calculated upon."

"I am too happy, sir, to have been the means of affording you any assistance. I am not a regularly trained sportsman, whatever you may think; but some encounters of this sort have happened to me before; so that, perhaps, I may say, '*Sono anch' io cacciatore.*'"

"You may say so, indeed, if it so pleases you."

All were now intent upon the deer, which was a first-rate one: he had few points to his horns, being one of those originally marked out as the fattest; he was beautifully cleaned, and all the operations being carefully performed, Tortoise thought it high time to satisfy his curiosity. He learned from the Sassenach that he was an artist, and travelled over the country, making sketches, with a light knapsack at his back; he had come that morning from Badenoch, and the Highlander before mentioned was his guide. He was a man, *factus ad unguem*, and a magnificent walker, and at once recognised by the hill-men as the painter who came to Blair two years before, and took Macintyre, with the Duke's permission, as his guide to Braemar forest. Now, Macintyre was one of the stoutest walkers in Atholl; no step was lighter or more elastic up the mountain,—none steadier or more **iron-like** when he bounded **down the** steep: to him was given strength, activity, **and** endurance of fatigue, beyond the common lot **of man**; he knew his superiority, and was proud of showing it; **but,** intent as he was in making a grand display **to astonish** the artist, he found himself totally discomfited. "The de'il was in the man; he skelped awa quite aisy, with a wee bit knapsack and umbrella to boot;" and although Mac very cannily slipped a **few stones into the** knapsack, he was beat the whole way; **and it was a laugh** against him to his dying day.

The artist having hinted that these sort of **encounters** had chanced to him before, Tortoise drew from him the following account of one of them:—

He had walked over Norway **on a sketching tour, and once joined a party of Norsemen who were ringing the bear.** He carried no fire-arms, he said, like the rest of the party, always preferring **close combat**;—nothing but his sketching stool. This, when produced, was found to be a circular piece of heavy oak timber, divided into three parts, fitting closely, **so as to** unite, and rivetted together in **the** centre; but when detached by a sort of twist, the extremities were spread,—the lower ones forming feet, and the upper ones a seat, by hitching some sort of sacking on **their** points. The thing is a sketching stool **in common use**,— his only differed from others by being **made of the most** solid oak, so that in good **hands** it was **a very effective** weapon; and it was with this **that he** had been attacking the stag.

"I was on skidor," said he, "**which you know is a sort of** long wooden skate, which enables you to get over the snow **at a** quick pace,—rather unmanageable, however, by a novice like myself. A young bear having been discovered **in** a cave, I begged he might be put at my discretion, and **that we** might have a combat *à l'outrance.* They talked **a** great deal of nonsense about danger, **but at** length the point was conceded. I roused the beast with a great stone, **which** hit him somewhere on the *os frontis.* Out came Bruin with a growl, and I then belaboured him over the head, and I really believe I should have **had the** best of it, being pretty expert at single-stick, could I have made any impression on the beast; but he only shook his head a little, as if he dissented from my conduct. He seemed much given to apathy—indeed I never saw a more phlegmatic animal; nevertheless he kept advancing upon me, and, at length, in spite of my blows, which were numerous and heavy, reared himself on his hind legs, and fairly got me within his foul hug. I assure you, upon my credit, I never felt more uncomfortable in my life; but the Norwegians, taking the alarm, ran in and dispatched him with their long knives: for this they received my forgiveness, though the combat was somewhat sullied, the rather, as I found the beast was powerful and resolutely inclined, though I would willingly have had a longer tussle with him. He is not a very terrible animal after all, but, on the contrary, somewhat too loving and close in his embraces, whereof I felt the effects for a considerable time afterwards.

"But, really, your Norwegian is always too hasty with his weapons. As an instance of what I say, I must tell you that I went with one of these barbarous huntsmen in quest of a salmon. Day after day, and week after week, did I toil without success; believe me, sir, in all that time I never saw a fin. At last the long-desired moment came, —I hooked a prodigious monster; the natives were astounded at his portentous size,—nay, some went so far as to say that he was no salmon, but the great sea-snake, called Jormungandr, in person, whom Thor fished for with a bull's head; but it proved to be a salmon after all, and not the great sea-snake.

"Soon after I hooked him he made a prodigious rush, which brought him on the channel in bare water; the officious Norwegian immediately tucked a large iron hook into him, which was fastened to the end of a long stick, and fairly hauled him ashore.

" Being extremely disappointed to find my sport terminate so suddenly, I obliged him to put the fish back into the river, that I might kill him *secundum artem*. This he was at length persuaded to do, though I must say he performed it with a very bad grace.

" The fish, once more in his element, began to exhibit

most astonishing power and activity, bending my rod like a willow wand, and making my arms quiver again; his runs were so strenuous and rapid, that one of my fingers coming in contact with the line, was deeply cut by it. After various manœuvres on his part (which I would fain hope I defeated with some degree of dexterity), he at length darted down the stream, and ran out nearly all my line; then he shot suddenly across the river, and went up under the opposite bank: I pulled strenuously, but my line seemed fixed to one particular spot; and whilst I was looking at that spot, where I conceived the monster to be, I just glimpsed him about twenty yards above, lunging out of the river, lashing his huge tail, and towing my tackle after him. Soon after this my line came up quite easily, and upon examination I found it about fifteen yards minus of its fair proportion. As for the salmon, I never saw or heard of him again."

"Aye! In Scotland this is what we call being drowned; meaning that the line is so, the action of the current and weight of water forming it into an immense curve, from which position it can with difficulty be extricated; but when you next hook any thing resembling Jormungandr, you had better endeavour to take the management into your own hands, and not suffer the snake or salmon, as it may be, to manage you; and if he runs out your line with a rush down the stream, follow, wind up, and keep above him; should he then attempt to cross, keep your line as short as you can, hold your rod aloft, and give him the butt. For if you once suffer him to cross to the opposite bank with so long a line as you appear to have had, he will not become your property—never shall you rejoice over his tinselled sides as he lies glittering on the pebbles. Some water-elf (for such, I am told, there are in Norway) never fails to interpose a great stone or rock between you and your fish; you toddle up the river all too late; and your tackle, assuming Hogarth's line of beauty, bears against this obstruction; the salmon pulling on one side against the concealed rock, and you unwittingly on the other; so that betwixt your united efforts, a fracture must inevitably take place, were your line even as strong as that used in trolling for the great water-bull of yore, when they baited with a sheep's head. My advice comes somewhat late, to be sure; but it may be of service to you hereafter.

"But you really came too late into the world, sir, and; should rather have flourished in the time of the Lapithæ I am convinced you would have been as wonderful as the best of them, at least the poets would have made you so, which, when a man is dead, you know, is the same thing; and, indeed, had you to-day advanced much closer in the combat with this dun beast, you might by this time have been a ghost, and taken your rank amongst the shades of Ossian's heroes. His horns stab fiercely, and when attacked he is altogether very redoubtable.

"Still I do homage to your wonderful activity, as well as to your gallant bearing: overtake you we could not, practised and trained as we are; though this may be in some measure accounted for from our previous exertions—the extent of which you will comprehend when I tell you that we brought this stag from yon mountain top, which you see melting into air in the extreme distance—and that from the said point to the place where we now stand, we have pulled up but twice, and that but for a brief space. We have had some sharp bursts, I promise you, which you have been pleased to extend: my friend, whom you see coming up, will bear witness to this. But really, now all is well over, I am much gratified at the pleasure you have received. We do not see such sets-to every day."

The wounded stag had by this awkward encounter taken the deer-stalkers so far out of their cast, that the day's sport was considered as ended. So the whiskey-bottle went round, and all were gossipping together like brothers.

The Highlander was a well-known good companion, pretty considerably addicted to poaching, like many of his compeers; but in this instance he well knew that he could not appropriate the deer, and that the rifleman must be in pursuit, so that he would willingly have stopped the stranger, had it been in his power to overtake him.

There was a great deal of merriment between the Atholl men and this Highlander, who was the Gown-cromb, or blacksmith, of some village in Badenoch. He was taxed, but in a merry mood, with many dexterous feats of poaching, and driving the duke's deer to the north, when the wind served, which he did not altogether deny.

"Well," said Tortoise, "take some more whiskey, and a pinch of snuff from my mull; but you must not steal the duke's deer, man"——

"Hout-tout! Ye're a true Sassenach, an' the like o' ye chiels aye ca'/liftin' stealin', which is na joost Christianlike."

"Well, what would you give for such bonny braes and birks and rivers as are in the forest of Atholl, if they could be transferred to your wild country?"

"And are there nae bonny braes and birks in Badenoch? Ye're joost as bad as our minister; but fat need the man say ony thing mair aboot the matter, fan I tell 'im that I'll prove, frae his ain Bible, ony day he likes, that the Liosmor, as we ca' the great garden in Gaelic, stood in its day joost far the muir o' Badenoch lies noo, an' in nae ither place aneth the sun; isna there an island in the Loch Lhinne that bears the name o' the Liosmor to this blessed day? fan I tell you that, an' that I hae seen the island mysel, fa can doot my word?"

"But, Mac, the Bible says the garden was planted eastward, in Eden."

"Hout! aye; but that disna say but the garden micht be in Badenoch! for Eden is a Gaelic word for a river, an' am shaire there's nae want o' them there; an' as for its bein' east o'er, that is, when Adam planted the Liosmor, he sat in a bonny bothan on a brae in Lochaber, an' nae doot lukit eastwar' to Badenoch, an' saw a'thing sproutin' an growin' atween im an' the sun fan it cam ripplin o'er the braes frae Atholl in the braw simmer mornings."

"But, Mac, the Bible further says, they took fig leaves and made themselves aprons; you cannot say that figs ever grew in Badenoch."

"Hout-tout! there's naebody can tell fat grew in Badenoch i' the days of the Liosmor; an' altho' nae figs grow noo, there's mony a bonny *fiag* runs yet o'er the braes o' both Badenoch and Lochaber. It was fiag's skins, an' no fig blades that they made claes o'. Fiag, I maun tell you, is

Lochaber Gaelic for a deer to this day; an' fan the auld guidman was getting his reproof for takin' an apple frae the guidwife, a' the beasties in Liosmor cam roon them, an' among the rest twa bonny raes; an' fan the guidman said, 'See how miserable we twa are left: there stands a' the bonny beasties weel clad in their ain hair, an' here we stand shame-faced and nakit—aweel, fan the twa raes heard that, they lap oot o' their skins, for very love to their sufferin' maister, as any true clansman wad do to this day. Fan the guidman saw this, he drew ae fiag's skin on her nainsel', an the tither o'er the guidwife: noo, let me tell ye, thae were the first kilts in the world."

"By this account, Mac, our first parents spoke Gaelic."

"An' fat ither had they to spake, tell me? Our minister says they spoke Hebrew; and fat's Hebrew but Gaelic, the warst o' Gaelic, let alane Welsh Gaelic."

"Well done, Mac; success to you and your Gaelic."

"Success to me an' my Gaelic! I tell ye that the Hieland Society, or Gaelic Society, or a' the societies in the world, canna ca' again' my Gaelic! nor the name or origin o' the first dress worn by man, for—

'Ere the laird cardit, or the lady span,
In fiags' skins their hale race ran.'"

"We would require proof for this, Mac."

"Proof, man! disna your Bible say, 'cursed is the ground for Adam's sake,' an' that curse lies on Badenoch an Lochaber to this day; for if there be in all Scotland a mair blastit poverty-stricken part than 'ither o' the twa, may Themus Mac-na-Toishach's auld een never see it! an' for the truth o' fat I'm saying, its joost as true as any story of the kind that's been tauld this mony a day: *let them contradic me fa can.*"

Thus the Gown-cromb's wit at length fairly got the better of his patriotism.

CHAPTER III.

Forests of Badenoch, their rights and divisions.—Legend of Prince Charles.—Cluny Macpherson.—Adventure with a wolf.—Macpherson of Braekally.—Children lost on a moor. — Sportsmen benighted. — Witchcraft. — Uncomfortable position.—Fraser's cairn.—Boundaries of Gawick.—Fate of Walter Cumming.—Wrath of a fairy.—Destructive avalanche.—Convivial resolution.—Arrival at Bruar Lodge during the night-storm.

> The sun went down behind the hill,
> The moor grew dim and stern;
> And soon an utter darkness fell
> O'er mountain, rock, and burn.

THE party now separated, the artist being bound for Blair. Tortoise and his friend struck across the hills towards Bruar Lodge, from which they were about eight or nine miles distant.

"Not bad, that supposition of our friend the artist," said Tortoise, "that he had hooked the great sea-snake; but one does hook strange things sometimes; as, for instance, Mr. James Rose, a friend of Mr. Skene of Rubislaw, was fishing on his property in the river Dee, it was snowing very thickly, and he had on his line a large fly, full four inches long, called there the black dog. In a short time he hooked what he conceived to be a fine strong salmon, who, however, worked as salmon never worked before, dragging the fisherman down the stream at the top of his speed, and making his arms quiver again; at length, to his great surprise, the animal began to give tongue, and he found he had hooked an otter by the muzzle. This increased his ardour, and he dashed along, at some risk, through the water, and over great blocks of stone, till at length a high projecting rock impeded his progress. Mr. Rose, however, was determined enough to throw himself into the Dee, and swim for some distance, rod in hand, after the otter; but unfortunately, his tackle failed, and the brute at length got off. Probably, however, he was killed afterwards; for a tenant of Mr. Skene, whose house was close to the water, was awakened one clear frosty night by screams and extraordinary sounds issuing from the river: he arose quickly, under an impression that some one had fallen into the Dee; when, to his relief, he descried two otters upon a large mass of floating ice, fighting for a salmon, which they had dragged upon it. They were screeching and yelling in fierce combat. The man loaded his gun, and fired at them with success; for when he arrived with his boat, he found one of the otters killed, and a beautiful salmon of twenty pounds beside him, with a piece only bit out of his throat; he got a good price for the otter's skin, and fed his family with the salmon.

"And now, as we are journeying on," said Tortoise, "I will endeavour to lighten the way by giving you a true description of the Badenoch country. I am putting together a short account of the principal forests in Scotland, and I meant to have reserved Badenoch for your perusal with the rest; but as you have just passed through a large tract of it,—and as the Gown-cromb rather libelled his own country, and, moreover, gave you but an apocryphal version of its history, I will take this opportunity of telling mine.

"The account I am about to relate, as well as I can from memory, was most obligingly given to me by Cluny Macpherson, chief of Clanchattan, a very celebrated and accomplished sportsman. Thus, then, it runs:—

"The Earls of Huntly possessed in former times by far the most extensive range of hills and deer forests in Great Britain; they commenced at Benavon, in Banffshire, and terminated at Ben-nivis, near Fort-William, a distance of about seventy miles without a break, with the exception of the small estate of Rothiemurcus, which is scarcely two miles in breadth where it intersects the forest.

"This immense tract of land was divided into seven distinct portions, each of which was given in charge to the most influential gentleman in its neighbourhood. The names of the divisions or forests were,—firstly, Benavon, in Banffshire secondly, Glenmore, including Cairngorm; thirdly, Brae-feshie; fourthly, Gaick;* fifthly, Drumnachder; sixthly, Benalder, including Farrow; and, lastly, Lochtreig, which extended from the Badenoch march to Ben-nevis; these are all in Inverness-shire.

"These divisions are very extensive; Benavon comprehends about twenty square miles, Glenmore the same

---

* Spelt also *Gawick*, and *Gaig*.

quantity, Brae-feshie about fifteen, Gaick about thirty, Drumnachder twenty-five, Benalder fifty, and Lochtreig sixty; in all about two hundred and twenty square miles.

"The whole of this vast tract was not solely appropriated for breeding deer, for tenants were allowed to erect shielings on the confines of the forest, and their cattle were permitted to pasture as far as they chose during the day, but they were bound to bring them back to the shielings in the evenings; and such as were left in the forest over night were liable to be poinded.

"These regulations answered very well between Huntly and his tenants, but they made an opening for small proprietors, who held in fee from the Gordon family, to make encroachments, and in course of time to acquire a property to which they had not the smallest legal title.

"In other respects, rights were more rigidly adhered to; for the old forest laws, which were exceedingly severe, were enforced to the utmost in this district; mutilation, and even death, were resorted to. It is upon record, that Donald of Keppoch hanged one of his own clan, in order to appease Cluny Macpherson for depredations committed in the forest of Benalder; and it is a known fact, that another person, called John Our (John the swarthy), had an eye put out, and his right arm amputated, for a similar offence; and it is also said, that he even killed deer afterwards, in that mutilated condition.

"No alteration took place in these forests till after the Rebellion of 1745, when the whole was let for grazing, with the exception of Gaick, which the Duke of Gordon continued as a deer forest until about the year 1788, when it was let as a sheep walk, and continued so until 1816, when the late Duke of Gordon (then Marquis of Huntly) re-established it; and it is now rented by Sir Joseph Radcliffe. But in consequence of cattle being admitted to summer grazing, the present number of deer, as I am informed, is not great; probably not more than between two or three hundred. The deer in this forest are small, and chiefly hinds; but, in all the other named forests, it was not uncommon to kill harts that weighed twenty-four stone, and even up to twenty-seven, imperial weight.

"The forest of Benalder is now rented by the Marquis of Abercorn, from Cluny Macpherson, chief of Clanchattan; but as the sheep were only turned off in 1836, there are not many deer in it as yet; still, as the Marquis of Breadalbane's forest is not far distant, they will, no doubt, accumulate rapidly under such excellent management.

"This forest lies on the north-west side of Loch Erroch, and contains an area of from thirty to thirty-five square miles: the position is in a south-west direction; the boundary on that side is the small river Alder; on the north-west it is limited by Beallach-na-dhu (the dark vale), and the river Coolroth (which signifies a narrow and rapid stream); and on the east it is bounded by Loch Pallag and the hill of Farrow.

"The mountains are lofty, probably near 4000 feet above the level of the sea, and many of them of picturesque character and majestic appearance. I must not omit, that there is a lake of two miles in circumference, at an elevation of at least 2500 feet, called Loch Beallach-a-Bhea (the Loch of the Birchin Gap). So much for the boundaries, extent, and character of this celebrated domain.

"The legends connected with this forest are numerous and interesting. In Benalder is a cave which gave shelter to Prince Charles Stuart for about three months after he made his escape from the Islands, where he so imprudently entangled himself. When he came to Benalder, he was in a most deplorable state—covered with rags and vermin; but there he was treated with kindness and hospitality; and during the period of his stay, he made considerable progress in the Gaelic language. Cluny Macpherson and Lochiel, faithful, high-minded, and loyal, were his constant companions; and they were attended by a few trusty Highlanders, who carried to him every necessary, and many of the luxuries of life.

"Cluny had generally the charge of this forest in olden times. On one occasion, a nephew of his, a young man, met a party of the Macgregors of Rannoch, who were upon a hunting excursion: there were six of them; but Macpherson, who had still a stronger party, demanded their arms: to this the Macgregor leader consented, with the exception of his own arms, which he declared should not be given to any but to Cluny in person. Macpherson, however, persisted in disarming the whole, and in the attempt to seize Macgregor, was shot dead upon the spot. The Macgregors immediately fled, and effected their escape: one alone suffered, who was wounded in the leg, and died from loss of blood.

"This unlucky circumstance was attended with no farther evil consequences—no lasting animosity—no secret vow of mutual extermination; but, contrary to usual custom, it had the effect of renewing an ancient treaty between the two clans, for mutual protection and support.

"When Cluny Macpherson resolved upon departing to France, on account of the share he had in the affair of 1745, he called upon a gentleman with whom he was intimate, and who was a noted deer-stalker (Mr. Macdonald of Tulloch), and said that he wished to kill one more hart before quitting his native country for ever: the proposal was cheerfully accepted by Macdonald, and they proceeded to Benalder accordingly.

"They soon discovered a solitary stag on the top of a mountain; but just as they had stalked almost within shot of him, he started off at full speed, and went on end for about two miles; he then stood for a few minutes, as if considering whether he had any real cause for alarm, and at length deliberately walked back to the very spot from which he first started, and was shot dead by Cluny. This circumstance was considered a good omen, and the prosperous interpretation was not falsified by future events.

"As for the forest of Glenmore, I would advise you to keep clear of it, unless, like the northern champions of old, you delight in encounters with military spectres; for it is said to be haunted by a fairy knight or spirit called Lhamdeargh, in the array of an ancient warrior, having a bloody hand, from which he takes his name. He challenges those he meets to do battle with him; and as lately as 1669 he fought with three brothers one after another, who immediately died thereafter.*

---

* Account of Strathspey, *apud* Macfarlane's MSS.

"I must now tell you of an adventure that happened to Mr. Macpherson of Braekally, when he had the charge of the forest of Benalder. He sallied forth one morning, as he was wont, in quest of venison, accompanied by his servant. In the course of their travel they found a wolf-den (a wolf being at that time by no means a rarity in the forest). Macpherson asked his servant whether he would prefer going into the den to destroy the cubs, or remain outside and guard against the approach of the old ones. The servant preferring an uncertain to a certain danger, said he would remain without; but here Sandy had miscalculated, for, to his great dismay, the dam came raging to the mouth of the cave; which, no sooner did he see, than he took to his heels incontinently, without even warning his master of the danger. Macpherson, however, being an active, resolute man, and expert at his weapons, succeeded in killing the old wolf as well as the cubs; and in coming out of the den espied his servant about a mile off, to whom he beckoned; and, with scarcely a remark upon his cowardly conduct, told him, that as it was now late, he intended to remain that night in a bothy at Dalenluncart, a little distance off. They accordingly proceeded to this bothy, and it was quite dark by the time they reached it.

"Macpherson, on putting his hand on the bed to procure dry heather for lighting a pipe, discovered a dead body; and without taking any notice of the circumstance, merely remarked,—'I don't like this bothy; we will proceed to Callaig, about a mile off, where we shall be better accommodated.' They accordingly went to this other bothy; and on arriving there, Macpherson pretended that he had left his powder-horn in the bothy they had just quitted, and desired his servant to go after it, telling him that he would find it upon the bed. The servant did as he was desired; but instead of finding the powder-horn, he placed his hand upon the dead man; which, to one of his poor nerves, was a terrible shock. He then hurried back in great agitation; and, on reaching the other bothy, found it, to his great dismay, dark and deserted, his master having set off homewards so soon as he had started for the powder-horn. Terrified beyond measure at this second event, he proceeded home, a distance of about twelve miles of dreary hill, where he arrived early in the morning; but the fright had nearly cost him his life, for he fell into a fever, and it was many weeks before he recovered.

"This Macpherson of **Braekally** was commonly called Callum Beg, or Little Malcolm; and there is reason to believe that he was one of those who fought in the famous battle of the Inch of Perth, in the reign of Robert the Third.

"An affecting circumstance happened in this district many years ago. Two children of tender age wandered from a neighbouring shieling in search of berries and wild flowers, and such pastime as innocent and happy souls delight in:—they never returned to their lonely dwelling; but after an anxious search, and a lapse of many days, were found dead, and locked in each other's arms. The place is still called Laggan-na-cloine-a-Caouch, or, the Hollow of the Affectionate Children.

"To recur to the deer, I must tell you, that it is confidently asserted that a white hind continued to be seen in Benalder for two hundred years; and there is at this present time a hind which was marked twenty years ago: she is well known to the shepherds, from the circumstance of both ears being cut off, which gives her an appearance too remarkable to be mistaken. There was also a large hart, well known in the forest for a period of thirty years;—he was said to carry eighteen branches. He has disappeared, however, during the last three years; but it has not been ascertained what has become of him,—whether he has been killed, died a natural death, or has changed his ground. There is now also a hart, which has been remarked for many years; he has a very peculiar formation of antlers, and it is well ascertained that he was shot through the body seven years ago, and is now perfectly recovered. I mention this chiefly to prove, from other evidence than my own, that a deer that has been wounded, has ever afterwards his horns deformed.*

"My story, I fear, has been a tedious one, but happily for

---

* Vide p. 29. Chap. I. on the Nature and Habits of Red Deer.

you I must now come to a stop, for all your attention will be required in picking your road; we have some very uncomfortable ground to pass over. Had the moon kept clear we might have made our way tolerably well, but that black cloud has completely mystified us."

In truth, it had become so impenetrably dark, that it was impossible to distinguish the nature of the moor,—whether the foot was to alight upon the top of the moss hag, or to sink down in the bog; the burns themselves, which ran silently, were not discernible,—no light from the sky being reflected on them. Each man struggled on as best he might; but the hill-men supported Lightfoot with that kind care and hospitable attention, which is the characteristic of every Highlander, from the highest to the lowest.

"Ye mun gang cannily, sir, an dinna pit yer fut doon rashly, for the bog is deep, it'll tak ye up to the weem; mony's the beast that has been lost in it. It was na lang sin' Sandy Macgregor, him that drives the cattle, lost his bonny cow,—the milk had been takken afore by some inveesable hand, or may be by the evil eye, and then the beast was gone a-the-gither. For twa days he lookit ower a' the green grazings, where aiblins she micht have strayed; aweel, on the third day, he saw the gathering of the ravens, and the waving of the wings, and the wheeling aboot in the air, and heard the hoarse croakings; and when he wun to the place, there was his bonny beast stuck fast, stark deid, a wee bit to the wast of whar yer honor stands the noo; and the foul birds had pickit out his een, and ate his flesh. They say that if Sandy had found the cow when the hide was fresh, and had takken it aff, and wrapped himsel in it, mony strange things would he have heard that nicht on the moor. Wha can say what thae birds may be, gin they were in their ain proper shape."

"All this is excellent comfort, my good friend; but why did you bring me here, to devote me to your bog kelpies,—do you wish to see another beastie lost, and food for the raven?"

"God bless you, sir, haud up, and dinna be afeard, ye shall no come to harm; tak my hand, and joost feel the moss a wee bit afore ye trust till it. Sandy, man, gang forrit a step or twa."

Sandy did as he was desired, and a loud splashing was almost immediately heard, like to the rising of a muckle salmon, when he attempts to spring up the falls of the Garry.

"Sandy, man, I'm thinking ye've got intill the burn."

"It's nae burn ava', it's joost a deep pool. Ye mun keep mair to the wast. Its fearfu' dark, and as sure as deid the evil spirit is abraid,—he couldna have harmed me in the burn, for you ken he has nae power in rinnin water. I am as weel acquent wi' this moss by day and by nicht as ony man in Atholl, and never pool was there here afore."

"In pool or ford can nane be smur'd,
Gin kelpie be nae there."

To describe the toil of the party through these bogs, pits, and moss hags, would be only to utter a repetition of the same disasters. The darkness was so deep that the men could not distinguish each other; and although their footsteps fell cautiously, yet not one of the party escaped continual floundering; the individual wrath and vexation was at first at a pretty high pitch; but with the exception of a slight exclamation or so, it was most philosophically suppressed. And when at length all were found to be in similar perplexity, there was more merriment than anger. Everything, however, whether sweet or bitter, has an end, and so at length had this their pilgrimage through the Slough of Despond.

As soon as they were fairly through, the blank moon, so coy when she was courted, shone out for a brief moment, and gave them a glimpse of a herd of deer just passing into the shadow. And now they came down to a burn, which, wet as they already were, they waded without hesitation. Lightfoot alone was carefully carried over on Fraser's back, for the channel was obstructed here and there by large blocks of granite, which the constant attrition of the water makes so slippery, that no unpractised person can step on them with security; and when he loses his footing (as lose it he must), down at once he goes into the deep hole that the current always excavates at their base. But the sinewy and well-shod Highlander went firmly and safely through with his burthen, the legs alone dangling in the water.

This portage was absolutely necessary, for our friend had the disadvantage of London shoes, which are somewhat of the neatest; and as the captain of Bewcastle said to Wat Tinlinn, the heels risp,* and the seams rive.†

They now came to firmer ground, and resolved, though it was somewhat out of their way, to strike across to the firm cart-track. This was so overgrown with heather, that it was not very distinguishable in day-time; and they were now only assured of their arrival at it by scraping with their feet, and thus ascertaining that the ground was hard.

"We are now at Fraser's cairn, and the Lord of Lovat's spirit may be abroad, calling for his horse. Are you not horribly afraid, Peter?"

"Hout-tout! Clish-ma-claver, I'm o'er auld farran to be fleyed for bogles."

"And now, Lightfoot, as our difficulties are fairly over, and you have your attention at liberty, I will finish my description of Badenoch, by giving you an account of its celebrated forest of Gawick. Should you like to hear it?"

"Very much; it will lighten our way; provided you will leave out everything that relates to bogs, burns, pits, and kelpies,—'an universe of death.'"

"Well, then, I must tell you that there are many very interesting circumstances connected with this forest; but though it may be somewhat dull, I will give you a description of its boundaries before I enter upon them.

"Its bearing is in a south-west direction; and it is bounded on the south by the hills 'of the braes of Atholl,' on the north and east by Glentromy and Corrybran, and on the west by the Glentruim and Drumnachter hills.

"In the centre of Gawick there is a plain about eight miles long, and in this plain there are three lakes—Lochandellich, Loch Bhroddin, and Lochindoune—all abounding with excellent char and trout. There is also another species of fish, called by the natives, Dormain. This fish is large, has a huge head, and is supposed to prevent salmon from ascending to the lakes: some of them weigh from twenty to thirty pounds. The hills on each side of this plain are

---

* Creak. † Tear.

remarkably steep, with very little rock, and of considerable altitude. On the western extremity there is a hill of a very striking appearance; its length is about a mile, its height about one thousand feet from the base of the plain; its shape resembles that of a house. This hill is called the Doune, and forms the southern limit of the forest. So much for the boundaries and locality; now for a tale of other times.

"Walter Cumming was killed by a fall from his horse in the forest of Gawick; he was the son, I believe, of one of the Cummings of Badenoch, and certainly a very profligate young fellow. Tradition says that he determined on making a number of young women shear stark naked in the farm of Ruthven, which was the residence of the Cummings in Badenoch. In the meantime he was called away on business in Atholl, and the day of his return was fixed for this infamous exhibition. When that day arrived, his horse galloped up to the court-yard, stained with soil and blood, with one of his master's legs alone hanging in the stirrup. Search was instantly made, and the mangled body of Cumming was found with two eagles preying upon it.

"This horrid circumstance was ascribed to witchcraft; and the eagles were supposed to be the mothers of two of the young girls intended for the shearing exhibition. The place where Walter was killed is called Leim-ramfian, or the Fingalian's Leap; and a terrible break-neck place it is.

"The fate of Walter is still proverbial in the Highlands; and when any of the common people are exasperated without the power of revenge, 'May the fate of Walter of Gawick overtake you,' is not an uncommon expression.

"The belief in 'spirits of a limited power and subordinate nature' dwelling amongst woods and mountains, is, as you know, common to all nations, and more particularly to such as are of a wild and romantic character. The lonely man who journeys over a vast uninhabited space, feels himself almost unconnected with human society; and when darkness falls upon the moor, objects of dubious form loom around him and disturb his imagination.

"Thus traditions of witches and fairies are numerous in

the forest of Gawick; one at least I will give you as a specimen of their character.

"Murdoch, a noted deer-stalker, went at sunrise into the forest, and, discovering some deer at a distance, he stalked till he came pretty near them, but not quite within shot. On looking over a knoll, he was astonished at seeing a number of little neat women, dressed in green, in the act of milking the hinds. These he knew at once to be fairies; one of them had a hank of green yarn thrown over her shoulder, and the hind she was milking made a grab at the yarn with her mouth and swallowed it. The irritable little fairy struck the hind with the band with which she had tied its hind legs, saying at the same time, 'May a dart from Murdoch's quiver pierce your side before night;' for the fairies, it seems were well apprised of Murdoch's skill in deer-killing. In the course of the day he killed a hind, and in taking out the entrails he found the identical green hank that he saw the deer swallow in the morning. This hank, it is said, was preserved for a long period as a testimony of the occurrence.

"This was not our deer-stalker's only adventure; for upon another occasion, in traversing the forest, he got within shot of a hind on the hill called the Doune, and took aim; but when about to fire, it was transformed into a young woman. He immediately took down his gun, and again it became a deer; he took aim again, and anon it was a woman; but on lowering his rifle, it became a deer a second time. At length he fired, and the animal fell in the actual shape of a deer. No sooner had he killed it than he felt overpowered with sleep; and having rolled himself in his plaid, he lay down on the heather: his repose was of short duration, for in a few minutes a loud cry was thundered in his ear, saying, 'Murdoch, Murdoch! you have this day slain the only maid in Doune.' Upon which Murdoch started and relinquished his spoil, saying, 'If I have killed her, you may eat her;' he then immediately quitted the forest as fast as his legs could carry him.

"This man was commonly called Munack Mach-Jan, or Murdoch, the son of John; his real name, however, was Macpherson; he had a son who took orders, and obtained a living in Ireland; and it is said that the late celebrated R. B. Sheridan was descended from one of his daughters.

"The most extraordinary superstition prevalent was that of the Liannan-Spell, or fairy sweethearts; and all inveterate deer-stalkers, who remained for nights, and even weeks, in the mountains, were understood to have formed such connexions. In these cases the natural wife was considered to be in great danger from the machinations of the fairy mistress.

"I now come to the relation of a story better vouched for, and of a melancholy nature, which happened in the year 1800. Captain John Macpherson, of Ballachroan, with four attendants, and several fine deer-hounds, was killed by an avalanche in Gawick. The house in which they slept (a strong one) was swept away from the very foundation, and part of the roof carried to the distance of a mile. This catastrophe was ascribed by some to supernatural agency, and a great deal of superstitious exaggeration was circulated, to the annoyance of Captain Macpherson's family and friends.

"But a more public, a more wide-spreading calamity, has lately befallen. The gallant spirit is fled—the benefactor, the father, the beloved of his people, is gathered to the tomb of his fathers. Mournfully has his lament sounded from the dumb heights of Corrie-arich, and been borne over many a mountain, and through many a glen, from the hospitable shores of the Spey to the dark pines of Rothiemurcus.

"Thus sadly ends my account* of the possessions of the former Earls of Huntly; and our journey is nearly ended also. Yon speck of light that you see at a distance below, about the size of a half-grown glow-worm, shines in Bruar Lodge. But let us mend our pace, for foul weather is coming on."

"Aye, you may mend yours, but you will mar mine: have at you, however. I am lighter than I was, and will be more frugal at breakfast another time; it was that which touched my wind. I must be eating venison pasty and mutton chops, forsooth; catch me at that again in the

---

* In allusion to the late Duke of Gordon.

morning. I'll match you yet. But by all the gods above, I will make such a dinner this night as shall content my inward man, and distress your menage exceedingly."

" Never fear, we are tolerably provided."

And now they were before the rugged walls of old Bruar. Out came a servant with a lighted candle, twinkling, and vainly contending with the rain and wind. The door at the end of the little passage opened upon a blazing fire of bog-wood and peat; the table-cloth was invitingly spread. Each before dressing drank a tumbler—

> " Di quel buon Claretto, benedetto,
> Che si spilla in Avignone."

And here we leave our men to the performance of such convivial deeds as Abernethy abhorred, and Cornaro was an utter stranger to.

CHAPTER IV.

TREATING OF THE NECESSARY QUALIFICATIONS FOR A DEER-STALKER, WITH A FEW HINTS TO HIM.

Necessary qualifications for a deer-stalker.—Curious attitudes required.—Sleep almost superfluous.—Advantages of baldness.—Self-possession indispensable.—Abstinence from drinking, and restrictions in food.—Gormandizer's pastime.—Royal diversion. —Sportsman's philosophy.—George Ritchie, the fiddler.—Crafty movements.— Currents of air.—Passing difficult ground.—Range of the rifle.—Firing at the target. Tempestuous winds.—A tyro's distress.—Overwhelming kindness.—Of speed and wind.—John Selwyn.—Wilson, the historian.—Glengarry.

> ——— " O, this life
> Is nobler than attending for a check ;
> Richer than doing nothing for a bauble ;
> Prouder, than rustling in unpaid-for silk."
> *Cymbeline.*

I WAS so impatient to get on the moor, and to plunge at once, as it were, in *medias res*, that I omitted in the first instance to describe what sort of properties a deer-stalker should be decorated with. And although most of these might be easily divined by the practised sportsman from a perusal of these pages, still it may be as well to touch slightly upon a few others that are absolutely indispensable.

If a sporting gentleman was asked what was the best make for speed and endurance of fatigue, he would probably describe his own figure as accurately as possible, and that with the greatest appearance of candour, looking around upon his fair or foul proportions, as it may happen. In this there is abundance of encouragement; and, indeed, I am inclined to think that men go in almost all shapes, excepting, perhaps, those of Geoffrey Hudson, Daniel Lambert, and the Irish or any other giant. One of the most active men I ever saw was Richmond, the black pugilist, and he was knock-kneed to a deformity. Set before me a man that is long from his hip downward, closely ribbed up, and with powerful loins; take care that he be straight, and of the happy medium between slim and stout; let his muscle be of marble, and his sinews of steel. Heavens, how the fellow will step out! And what tremendous odds are half a foot in every step! See with what an elastic spring he recovers his legs! I swear by Atalanta and Achilles, the swift of foot, that this is the man I would back to go right up the Andes without deviating an iota from the straight line. I must add, however, that his lungs should be pre-eminent, because in long runs (say of six or seven miles at a stretch), through bogs and over mountains, wind will be found an article most particularly in demand. After all, a man should be trained in the way he should go as soon as he is out of petticoats; if not, the symmetry of the Antinous will avail him nought. I have not the slightest doubt, indeed, but that Pan would have caught Daphne much sooner than Apollo. He would have made a much better run, and probably a better thing of it altogether.

Now, this is all very well; but your consummate deerstalker should not only be able to run like an antelope, and breathe like the trade winds, but should also be enriched with various other undeniable qualifications. As, for instance, he should be able to run in a stooping position, at a greyhound pace, with his back parallel to the ground, and his face within an inch of it, for miles together. He should take a singular pleasure in threading the seams of a bog, or in gliding down a burn, *ventre à terre*, like that insinuating animal the eel,—accomplished he should be in skilfully squeezing his clothes after this operation, to make all comfortable. Strong and pliant in the ankle, he should most indubitably be; since in running swiftly down precipices, picturesquely adorned with sharp-edged, angular vindictive stones, his feet will unadvisedly get into awkward cavities, and curious positions;—thus, if his legs are devoid of the faculty of breaking, so much the better,—he has an evident advantage over the fragile man. He should rejoice in wading through torrents, and be able to stand firmly on water-worn stones, unconscious of the action of the current; or if by fickle fortune the waves should be too powerful for him, when he loses his balance, and goes floating away upon his back (for if he has any tact, or sense of the picturesque, it is presumed he will fall backwards), he should raise his rifle aloft in the air, Marmion fashion, lest his powder should get wet, and his day's sport come suddenly to an end. A few weeks' practice in the Tilt will make him quite *au fait* at this. We would recommend him to try the thing in a spate, during a refreshing north wind, which is adverse to deer-stalking; thus no day will be lost pending his education. To swim he should not be able, because there would be no merit in saving himself by such a paltry subterfuge; neither should he permit himself to be drowned, because we have an affection for him, and moreover it is very cowardly to die.

As for sleep, he should be almost a stranger to it, activity being the great requisite; and if a man gets into the slothful habit of lying a-bed for five or six hours at a time, I should be glad to know what he is fit for in any other situation? Lest, however, we should be thought too niggardly in this matter, we will allow him to doze occasionally from about midnight till half-past three in the morning. Our man is thus properly refreshed, and we retain our character for liberality.

Steady, very steady, should his hand be, and at times wholly without a pulse. Hyacinthine curls are a very graceful ornament to the head, and accordingly they have been poetically treated of; but we value not grace in our shooting jacket, and infinitely prefer seeing our man, like

Dante's **Frati**, "*che non hanno coperchio piloso al capo;*" because **the** greater the distance from the eye to the extreme **point of the** head, so much the quicker will the deer discover **their enemy,** than he will discover them. His pinnacle **or** predominant, therefore, should not **be ornamented with a** high finial or tuft. Indeed, the less hair he has upon it the **better.** It is lamentable **to think that there are so few** people who will take **disinterested** advice upon **this or any** other subject; but **without** pressing the affair disagreeably, I leave it to **a** deer-stalker's own good sense to consider whether it would not be infinitely better for him to **shave** the crown of his head at once, than to run the **risk of losing** a single shot during the entire season. **A man so shorn,** with **the addition of** a little bog earth rubbed **scientifically over the crown of** his **head,** would be an absolute **Ulysses on the moor, and** (*cæteris paribus*) perfectly invincible. **Do this or not, as you please, gentleman;** I am **far from insisting upon it with vigour, because, to my utter shame and confusion, be it spoken, I never did it myself.**

When **Sir Francis Head fled** over the Pampas, mounted upon wild horses, as if upon the griffin of Astolfo, he must have felt a sense of buoyancy and freedom that it would **be** difficult to describe. Astride upon the monstrous crocodile, Mr. Waterton must have rejoiced in his novel position and fair feats of jockeyship. **But neither** Mr. Waterton, nor he the subduer of **the crocodile and python, can** possibly **feel more secret exultation than the well trained pedestrian,** confident in his speed, secure in his aim, and unbaffled in his science.

As to mental endowments, your sportsman **should have** the qualifications of an Ulysses and a **Phillidor combined.** Wary and **circumspect,** never **going** rashly **to work, but** surveying all his **ground** accurately before **he commences** operations, and previously calculating all **his chances both of** success and of failure. Patient under suspense **and disappointment,** calm and unruffled **in moments of** intense **interest,** whether fortune seems to smile or frown **on his** exertions; and if his bosom must throb at such times, when hopes and **fears** by turns assail it, he should at all events keep such sensations under rigid **control, not suffering them** to interfere with his equanimity, or to disturb the coolness and self-possession which at such moments are more than ever necessary to his operations.

And that he may preserve in all their due vigour and steadiness these indispensable qualities, he should add to them in his hours of leisure and refreshment the further graces of temperance and moderation. And here condemn me not, ye joyous editors of Maga, if I restrict my stalker to moderate libations after his toil.

Odogherty, be merciful; Christopher, put down thy bristles; for lo, I will not limit him as Sir Humphry does his fisherman, to the philosopher's half-pint of claret; but, if he exceed it, 'tis at his own peril. Wine and poetry go joyously together. Bacchus and Apollo were aye boon companions; but I never heard of Diana having attached herself to the jolly god, or of an amour between Hebe and Adonis. Hard work upon wine will parch up the body, and make the hand ricketty—you ken that yoursel', Christopher. A keen deer-stalker's walk will keep a horse in a pretty decent trot, and his run changes that trot into a gallop, a sort of eclipse pace. Would you then have him *Bacchi plenus?* Yes, I verily believe you would. Well, my good Anacreon, only just try that system yoursel' a wee bit. During the first week, your mouth will be drinking bog-water in every black pool you can find; in the next, your flesh will vanish from your solitary bones; and, in the third—yes, in the third, at latest—you will die by spontaneous combustion.

The best part of a bottle of champagne may be allowed at dinner: this is not only venial, but salutary. A few tumblers of brandy and soda-water are greatly to be commended, for they are cooling. Whiskey cannot reasonably be objected to, for it is an absolute necessary, and does not come under the name of intemperance, but rather, as Dogberry says, or ought to say, "it comes by nature." Ginger beer I hold to be a dropsical, insufficient, and unmanly beverage; I pray you avoid it; and as for your magnums and pottle-deep potations, why, really at this season of the year, as Captain Bobadil says, "We cannot extend thus far." When the nerves are unsteady, the rifle in the sportsman's man begins to betray a want of fixed purpose and resolution; it does, as it were, vibrate considerably. Under these circumstances the balls are apt to take many untoward directions, such as are wholly unlooked for, and not fitted to maintain his reputation. Very wanton courses they will sometimes take, dabbing into a bog, or smacking against a stone; the deer all the while scampering and galloping away, freedom in their air, and independence in their heels! Already they have broken out of your cast—now they vanish over the hill—and by the direction they are taking, it grieves me to say that you are not likely to see them again this blessed day.

Having thus somewhat stinted my rifleman in his potations, it may possibly be inferred that I allow him to make up for such abstinence in the article of substantial food. This is a great mistake; I permit him to do no such thing; and most particularly do I restrict him at breakfast.

Should a deer-stalker eat and stuff?—should he pamper the inward man? Shade of Abernethy forbid! He should go forth lank and lean like a greyhound; the most that can be permitted him is a few cups of coffee, a moderate allowance of fine flowery pekoe, some venison pasty, mutton-chops (both are easy of digestion), a broiled grouse, of course, hot rolls, dry toast, and household bread, with a few grapes to cool him. Peaches and nectarines may be put in his pocket, because, as he will be sure to sit upon them, they will do him no earthly harm, but rather confer a benefit by moistening the outward man. But here I must stop: at this point the muzzle must positively be put on; for would you have me fill my man with Findon haddocks, and all the trashy and unprofitable varieties of marmalade: red, green, and yellow? What a proposition! Oh, no; I say again in no manner, and by no means will I let him gormandise. After the slender fare above mentioned, he will bound along like a Grimaldi; and let me see a hearty eater that has the least chance with him.

Can a man with a full stomach dash up Ben Derig? Vain hope! He would sink down gently in the first bog; nought save his head appearing above the surface; and the raven would feed upon his scalp, as Ugolino did upon the cruel archbishop's.

Ye who eat long like your mothers, and fast like your fathers—ye, believe me, had much better remain at home with your household gods, and cultivate decisive apoplexies. Everybody will tell you how well you look; so let out your waistcoats and your waistbands most amply, my much cherished friends—eat, drink, and be happy; or if the god of sport be warm within you, if so great—such an inextinguishable ardour burns in your bosoms, arrange yourselves, I pray you, in an ample punt on a domestic fishpond, with a rod, a line, and that admirable contrivance the float; but let not your obese fingers aspire to dally with a rifle.

Tell me now, could you hit any given acre of land at fifty paces? I should rather think not. As for a rifle, then, have nothing to do with it, I beseech you, my good fellows, lest it should go off unadvisedly. We are ready to give you every possible credit for your private and domestic virtues;—you are good fathers, the best of husbands, and the most excellent of friends—in short, ornaments to society; much more valuable members of it, indeed, than we minions of the mountains. What! does not this satisfy you? Do you mutiny in your punt, and are you determined to reject our wholesome advice? Well, then, we admire your spirit which soars so high above your corporal capacity, and since you are so determined, we will grant you our license to sport with the stag after the self-same fashion with Queen Elizabeth.

Thus it was:—When the said Queen of glorious memory visited Lord Montacute at Cowdrey in Sussex, on the Monday, August 17th, 1591, her Highness took horse and rode into the park at eight o'clock in the morning, "where was a delicate bowre prepared, under which were her Highness' musicians placed; and a cross-bow, by a nymph with a sweet song, was delivered into her hands to shoot at the deere; about some thirty were put into a paddock, of which number she killed three or four, and the Countess of Kildare, one."*

---

* Nicoll's Progresses, vol. ii.

This is the exact thing for you, and I pray you not to omit the nymph with the sweet song.

After all, we doubt not your resolution to attack the stag, or any other fierce animal, for we have had a very high opinion of the courage of a well-fed man ever since we heard the story that Wilkes delighted to tell of Alderman Sawbridge, which, for your satisfaction, we will recount.

The Alderman was induced to go a-hunting, a sport that was novel to him; and having some sort of indistinct idea that danger was connected with it, he went forth in the uniform of the city train bands, to which he belonged. Being told that the hare was coming his way, he boldly laid his hand on the hilt of his sword, and replied, with perfect self-possession, " Is he sir? let him come!"

And now a word of advice to your well qualified sportsman—I beseech you, good sir, to bear bad weather and inauspicious winds with imperturbed philosophy. When the adverse day comes, as come it will; when the dark clouds gather round your desolate cottage, and the rain comes lashing and hissing along the moor, and the heather is uprooted by the blast, do not give way to despondency; but rest your toil-worn limbs, and be thankful that you have fire and shelter. Sit you down with your hand in your mantle (that is, your plaid), with the composed dignity of Aristides.

It is totally unavailing to look sulky, and to pace up and down the room, exclaiming at every step you take, What horrid weather! how very provoking! I never knew this sort of thing have the least effect upon the elements: betray not, I beseech you, the impotency of Xerxes, but fall back upon your resources. Read some amusing or instructive book, or if a book is apt to draw you to sleep (as it does full many a sportsman), get a piece of canvas nicely prepared by Mr. Browne of High Holborn, and paint your men and your dogs if you can; if you cannot, why then clean the locks of your rifles, sort your fishing tackle, and make flies; or if you are of a self-complacent character, you may summon your hill-men, and make it out, not in direct terms (you know how to manage it, I dare say), but by skilful inference, that you are, out of sight, the best shot in Great Britain: pass round the whiskey, and you may be certain of a ready acquiescence.

Then when the night closes in, you may call in George Ritchie, the fiddler and wit, if he happens to be in your train. Oh! George, how well I remember your speaking countenance—your capacious mouth—and your mighty ears. You are a good fellow, George, and were a most admirable deer driver to the lord of the forest, and for this I honour thee; but thou didst play me many a slippery trick by neglecting orders when thou wert wont to carry home the dead deer; for, instead of coming in behind my cast as instructed, thou didst ever cut in before me, and disturb all the ground in a most unsportsmanlike manner; and this thou didst transact most cannily, winding up a hollow with thy sheltie, that mine eye might not visit thee; yet I kent well enough what was going on, George, by the movement on the moor; but alas! poor George, you were growing old, and had a right to favour yourself a little; and then thou wert merry in hall, and thy quaint attitudes, and quainter countenance (whilst thou didst worry the strings of thy fiddle) did set the gillies in a roar:—for these, thy most excellent qualities, I do recommend thy presence to get up a Highland reel in a stormy evening.

I must now revert to you few, O happy mortals, "*quos æquus amavit Jupiter*," and I must candidly tell you, that I cannot turn you loose on the mountains to go rambling after your own inventions.

Enthusiasm you have, no doubt, else wherefore soar you to the mountain top? But this solitary qualification, indispensable as it is, will not set you up entirely. You must have extreme caution in certain situations, and at the same time, prompt decision and execution: boldness also, amounting to rashness in others; always, indeed, a happy mixture of the two in the same movement;—in short, you should be constituted something after the fashion of Sardus Tigellius—

> "Sæpe velut qui
> Currebat, fugiens hostem, persæpe velut qui
> Junonis sacra ferret."

I know nothing more beautiful than the running of a skilful deer-stalker, when the harts are in quick motion. He dashes after, or parallel to them, in order to come in at certain places; but never blindly, never straight forward, as if he could overtake them; but winding, sweeping, and lurching behind the ridges and hillocks, or down a narrow chasm, or up the stony channel of a burn, just keeping sight of the points of their horns; stooping or rising, moderating or increasing his pace according to circumstances, always preserving the wind, and taking care never to commit himself by coming upon such an open tract of ground as would fairly expose him to view; such blind rashness would hurry on the herd, and give them a fresh start for miles; for even if he should discover a solitary hillock, or block of granite, behind which he could find time to conceal himself for the moment, still he could not advance from this position, and he would be what is technically called "locked in."

Every person, I believe, who carries a rifle, is aware that when deer are disturbed, they always move up wind. They have an astonishing faculty of smelling the taint in the air at an almost inconceivable distance; being thus warned by instinct, they are enabled to avoid an enemy in front, and can go boldly forward over rugged ground and high points, without being surprised by an ambush. It would appear, then, at the first glance, that one's manœuvres, so far as relates to the wind, would be simple and easily conducted; but this is by no means the case,—the currents of air change according to the disposition of the ground; there are corries so situated that the swells of wind come occasionally from various quarters, and there are burns whose general tendency is in a direct line, but in whose various curvatures, the wind comes sometimes from the north, and at others from the opposite quarter; for it must be noted, that it always blows up or down a glen,—never across it.

Thus, in particular situations, you cannot ascertain the exact course of the wind without consulting that of the clouds, to which a hill-man always looks; but in all doubtful points, when the sky is cloudless, and the air tolerably still, a little tow dropped from your hand will indicate its course. When a lesser glen or burn debouches into another where the deer are on foot, and the current of air is one point only against you, your wind will be carried down the glen you pass, into the other at right angles to it, so that you must let all the deer pass the point of connection between the two glens before you cross the one in question.

It is impossible to describe the various nice points and wanes of the air that may occur in the course of the day; they can only be understood by long practice and observation; and observe, my good friend, that the most extreme caution is indispensable as to this point; for, without meaning any disrespect to you, you have such a *mauvaise odeur* about you, that the deer fancy you more formidable than you are, and your taint will make them break out of your cast: look not after them, I beseech you: it is vanity. By the help of a good pair of wings you may possibly fly; Icarus and the Ulm tailor did so before you; but those deer shall you never command on that inauspicious day.

The hill-men who act at a distance from you must use the same precaution with yourself in paying attention to the wind, and shifting their ground in obedience to any change that may take place in the course of the day.

There are some few cases when deer may be made to go contrary to their usual custom; thus in the forest of Atholl, when a herd comes out of Glen Croinie (which is a preserve, and may be called their home), they can readily be got back by good management, even if the wind is unfavourable — especially towards the evening, when they seek the pastures.

Deer may likewise be got down wind by sending men to take concealed positions in their front; the taint in the air will then turn them.

When there is a long line of deer on foot, in running parallel to them you must be careful not to get too forward, lest the tail ones get your wind; if, indeed, the deer have been pressed forward for a long distance, and are at all fatigued, it would not be prudent to do so at any rate, as in that case the fattest and heaviest harts always come lumbering behind.

When you discover deer with the glass at a considerable

distance, you may often approach the desired points without the necessity of being concealed by inequalities of ground. At what particular distance they will see you, must depend upon the state of the atmosphere and the nature of the ground you are traversing. If the point is dubious, you should always select the dark heather and bog to walk upon, and avoid the green sward, where you will be more easily descried. Be careful to expose as small a front as possible, walking rank and file, each file covering the leading one. Sometimes it happens that there is a small space only to be passed, in which you will evidently be visible; and in this case it is very difficult to elude the vigilance of the sentinels of the moor. The best way is to watch your opportunity when all are browsing, and then dart forward rapidly with your bodies bent across this dangerous point, one behind the other, as before described. I have often done this successfully; but it is a ticklish business, and will never succeed when you are near the quarry.

In all cases of approach, when it is necessary to advance in a stooping position, or to crawl, you had better keep a constant eye upon the men in the rear, for, believe me, no man is implicitly to be trusted; one will most unconscionably put his head up because, forsooth, his back aches insupportably; another likes to have a peep at the deer; a third (and he is the most unpardonable of all) does not like to have the burn water enter the bosom of his shirt, which is very inconsiderate, as nothing tends to keep a man more cool and comfortable than a well-applied streamlet of this description. So, look back constantly to the rear, that every gilly may do his duty, and observe that no man has a right to see the deer in approaching to get a quiet shot, except the stalker. In fact, after a certain distance is gained, no one but he and his Achates, who holds the spare rifles, should come forward at all.

The most perfect shots and celebrated sportsmen never succeed in killing deer without practice; indeed, at first, they are quite sure to miss the fairest running shots. This arises, I think, from their firing at distances to which they have been wholly unaccustomed, and is no reflection upon their skill. It is seldom that you fire at a less distance than a hundred yards, and this is as near as I would wish to get. The usual range will be between this and two hundred yards, beyond which, as a general rule, I never think it prudent to fire, lest I should hit the wrong animal—though deer may be killed at a much greater distance.

Now the sportsman who has been accustomed to shot guns, is apt to fire with the same sort of aim that he takes at a grouse or any other common game; thus, he invariably fires behind the quarry; for he does not consider that the ball, having three, four, or perhaps five times the distance to travel that his shot has, will not arrive at its destination nearly so soon; consequently, in a cross shot, he must keep his rifle more in advance. The exact degree (as he well knows) will depend upon the pace and remoteness of the object.

Deer go much faster than they appear to do, and their pace is not uniform like the flying of a bird; but they pitch in running, and this pitch must be calculated upon.

Firing at a target is a very necessary practice in the first instance, partly to gain steadiness and confidence, but principally to ascertain the shooting of your rifles at all distances. You can make no use of a change of elevation in your sights when deer are running; the best way, therefore, is to have one sight alone slightly elevated, the less the better, and to make the variation depend upon your aim. Having once become a fair shot at the target, I would advise no one to continue the practice. It is apt to make one slow and indecisive. One step often brings you into sight of the deer, consequently one spring makes them vanish from it, so that you must frequently take snap shots. Indeed, it is quite wonderful (as any experienced person can bear witness) how suddenly and unexpectedly they disappear, either by sinking under a hill, or running amongst the deep channels of a moss, or by a hundred means of concealment that the rugged nature of the ground affords them.

In firing down hill you must be very careful to keep your face low down to the sight, which sportsmen do not pay sufficient attention to; and think, therefore, that the ball mounts, which is a great mistake. When your head is too high, the line of vision does not follow the line of the barrel, but crosses it, and has a downward tendency, whilst the barrel perseveres in a more horizontal direction: and this is the doctrine of elevated sights.

You will often have to stop suddenly, and fire in the midst of a sharp run; or when you are dead blown; stand as steadily as you can, and be at once collected; practice alone can give you this power; and it will give it, for I myself was as sure at these sort of shots as at any other, provided the deer were running. I found it more difficult to take a quiet shot while lying on my stomach in the heather.

Sometimes the wind is so tempestuous that you have no power over the direction of your rifle. There are no means to counteract this, and you had better go home; but if it be not too violent, you can kneel on one knee, and get a rest by supporting your left elbow on the other.

Take care that the ramrods to your rifles be large and strong; they will otherwise be broken in the hurry of loading. I recommend you, moreover, to make one of your hill-men carry a very long and stout one in his hand, having a mark made in it at the length of your barrel, that you may ascertain the exact load. I used no other when this was at hand.

As for the sport itself, that no one can have a proper perception of till he is chief in command, and able to stalk the deer himself; and this he cannot do without long practice, close observation, and a thorough knowledge of the ground and habits of the animal. As an instance of this, one of the best shots in a rifle regiment was appointed some years ago to the office of forester in the Ben-Ormin Forest, in Sutherland; but being a stranger to the country, devoid of assistance, and without the means of good instruction in the craft, he was only able to kill one hart during two years of apprenticeship, and at length resigned his situation in despair. Novices, therefore, have necessarily a deer-stalker allotted to them from the forest, who very properly keeps the devoted rifleman in due subjection; he will not permit him to show a hair of his head above the heather on certain ticklish occasions, and the miserable youth is always totally unconscious of what is going on; he creeps and meanders through the black and miry channels of a bog, quite ignorant of the dire necessity for such a pastime; lies down to hand like a pointer, and runs till he is as breathless as an immerged oyster diver, he knows not why or wherefore. Thus the wretched felicity-hunter follows as best he may—

> "O'er rocks, caves, fens, bogs, dens, and shades of death,
> A universe of death."

One while his leg is wedged in amongst tenacious stony fragments, and at another he comes suddenly upon a deep chasm that fills his soul with unaffected apprehension. Meanwhile the deer-stalker goes on at a persevering killing pace, saying, "This way, this way, sir;" and never looking behind him to ascertain whether his patient is in his good ground or not; his words die away amongst the winds, and never reach mortal ear. Then, behold, when the deer come suddenly in view, he tells the staggering and breathless sportsman to shoot, always running forward himself, and placing his proper body (to say nothing of his flowing kilt) precisely in a mathematical straight line between the rifle and the harts, which he expects you to kill. Pleasant this to you; and, if in the excitement of the moment you obey his command, awkward enough for him! In getting a quiet shot, things may possibly be managed better, as to one part of the transaction; for if your adjutant will place himself between you and the deer (as right certainly he will), you may pull him back by the heel; or if you be not sufficiently powerful to make a good drag of him, you may admonish him in a friendly way, by a gentle insinuation of your gun-picker into the calf of his leg. You are not permitted to speak; and what else can you do?

You must by no means conclude, however, that your attendant means you anything but the most cordial kindness,—his zeal and fidelity in favour of those whom he has charge of is his great object; he means to take care of you as if you were his only son,—the remaining prop of his family. Anxious to give you every possible chance, he creeps, runs, and wades,—unmindful only that he is a son

of the mountains, whilst you, perhaps, were born in the Lincolnshire fens,—that his is the speed of the roe, and yours the pace of a frog; thus, whilst you are in such an exhausted state as to require the kindest and most unremitting attentions of the humane society, he is perfectly convinced that you are enjoying the highest degree of human felicity, unbroken in wind, and undecayed in strength.

In this dilemma what is to be done? I agree with you, that it is a thousand pities so fine a youth should perish prematurely; still I cannot allow you to speak of your distress; though that, indeed, you could not conveniently do, for want of breath, and if you could, you would only frighten the deer, without bettering your own condition.

You are at your last gasp, that is evident: perhaps, then, you had better do as the fat knight did, when the hot and termagant Scot was about to pay him "Scot and lot too," namely, to fall prostrate, and feign to be extinct, leaving Donald to speak a dirge over you in his most harmonious Gaelic.

"Death has not slain so fat a deer to-day."

Now, after all this, perhaps you will tell me that I have undervalued your powers. I dare say I have; there is not, indeed, the least doubt of it. To speak fairly, I think our young sportsmen from the south (I mean the most active of them) are fully as quick, and perhaps more so, than a Highlander, for a short distance; but when it comes to a trial of wind and endurance, your well-built, sinewy native will generally be found to be the best man.

In times of yore, however, we Sassenachs have produced huntsmen able and skilful in killing the stag. Not to mention the feats of Robin Hood and Little John, or the other unlicensed deer slayers "of merrie Sherwood," we are told that, "In the reign of Queen Elizabeth, John Selwyn, under-keeper at the park at Oatlands, in Surrey, was extremely famous for his strength, agility, and skill in horsemanship; specimens of which he exhibited before the Queen at a grand stag hunt at that park; where, attending, as was the duty of his office, he, in the heat of the chase, suddenly leaped from his horse upon the back of the stag (both running at the same time at their utmost speed), and not only kept his seat gracefully, in spite of every effort of the affrighted beast, but drawing his sword, with it guided him towards the Queen, and coming near her presence, plunged it in his throat, so that the animal fell dead at her feet. This was thought sufficiently wonderful to be chronicled on his monument, which is still to be seen in the chancel of the church of Walton upon Thames, in the county of Surrey. He is there represented on an engraved brass-plate, sitting on the back of a deer at full gallop, and at the same time stabbing him in the neck with his sword."*

This feat of John Selwyn has been paralleled very lately by one recorded in another page of this work; and in still earlier days, perhaps, was equalled in jockeyship, by Merlin Sylvester, the Wild, as mentioned by Geoffery of Monmouth.

"Merlin had fled to the forest in a state of distraction; and looking upon the stars one clear evening, he discovered, from his astrological knowledge, that his wife Guendolen had resolved upon the next morning to take another husband. As he had presaged to her that this would happen, and had promised her a nuptial gift (cautioning her, however, to keep the bridegroom out of his sight), he now resolved to make good his word. Accordingly, he collected all the stags and lesser game in the neighbourhood, and having seated himself upon a hart, drove the herd before him to the capital of Cumberland, where Guendolen resided; but her lover's curiosity leading him to inspect too nearly this extraordinary cavalcade, Merlin's rage was awakened, and he slew him with the stroke of an antler of the stag."

Formerly, it seems, the hunters went to the chase armed at all points, like the redoubted Alderman Sawbridge. Wilson, the historian, records an escape that befel him in the hazardous sport, whilst a youth and a follower of the Earl of Essex.

"Sir Peter Lee, of Lime, in Cheshire, invited my lord, one summer, to hunt the stagg; and having a great stagg in

---

* Antiquarian Repertory.

chase, and many gentlemen in pursuit, the stagg **took Soyle**; and divers, whereof I was one, alighted, and stood, **with** swordes drawn, to have a cut at **him at** his coming out of **the water**; the stagg then being wonderfully fierce and dangerous, made us youths more eager **to be** at him, but **he escaped** us all; and it was my misfortune **to** be hindered of my coming nere him **(the** way being sliperie) by a fall; which gave occasion to some, who did not know mee, to speak as if I had falne **for fear; which** being told mee, I left the stagg, and followed that gentleman, who first spake it; but I found him of that cold temper, that it seems his words made an escape from him, as by his denial and repentance it appeared.

" But this made mee more violent in pursuit of the stagg, to recover my reputation; and I happened to be the only horseman in, when the dogs sett him up at bay, and approaching nere him on horsebacke, hee broke through the dogs, and run at mee, and tore my horse's side with his hornes, close by my thigh. Then I quitted my horse, and grew more cunning (for the dogs had set him up again); stealing behind him, with my sworde I cut his ham-strings, and then got upon his back, and cut his throat; which as I was doing the company came in, and blamed my rashness for running such a hazard." *

Rashness! what rashness? Here's a fellow **for you now**; armed with a long sword, and probably in the uniform of the city **train bands**, he sneaks behind a stag at bay with fifty hounds fighting at his front,—fifty hounds and an historian are fearful odds. He then cuts his ham-strings in a dastardly manner, and puts an end to the brave animal's existence without doubt, by poking the end of his toledo, as Master Mathew has it, into the point of junction between the head and neck, precisely in the same manner in which the Laps kill their domesticated reindeer. As for his cutting the throat, I do not believe a word of it; he was of too cold a temper, and did no such thing, depend upon it,— he dared not so much as to look at his throat, being too fearful of his own.

---

* Quoted in the notes to the " Lady of the Lake."

This, however, was all mighty well for a young historian. We blame not his caution. We are circumspect ourselves. But we object to his bragging,—most decidedly object to it. The whole affair was a paltry one. Thy histories, great shade, I never read :—they may live like the great pyramids, or go to the tomb of all the Capulets,—it imports me not, —but shame on thy bragging of such a deed; shame on thee, I say, " thou chronicler of small beer."

Not thus was the bearing of the stout Glengarry, when he confronted the stag in the rugged pass of Glendulachan. Setting at nought the red glance of his eye, and unappalled by his tremendous means of defence, in rushed the gallant chief full at his front, and buried the sharp skene-dhu in his chest.

CHAPTER V.

A Scotch mist.—Visions of auld lang syne.—Retrospect.—The mist clears.—How to carry the spare rifles.—Storm in the mountains.—Sportsmen struck by a thunderbolt.—Willie Robertson's lament.—Macintyre's death.—Deer seen on the move.—Vamped-up courage.—Making a dash.—Unexpected success.—Dogs fighting.

> Stay, huntsman, stay; a lurid gloom
>   Hangs threatening o'er your head;
> The rain comes lashing o'er the moor,
>   The thunderbolt is sped.
>
> And mirk and mirker grows the hill,
>   And fiercer sweeps the blast:
> The heavens declare His wondrous power
>   Who made the mountains fast.

THE night has been dark and stormy, and the morning broke over the mountains in flames of red and amber; thin wreaths of mist were ascending from the Vale of Tay, and went twisting and flickering up the hill sides; there were no dark frowning clouds in the sky, but a sort of aqueous appearance about the light itself, that occasioned certain gloomy forebodings in the breast of our sportsman. True it is that he passed rapidly over the moor, as he was wont, and ate his usual slender meal with tolerable resignation. But to say that he enjoyed any thing like elevation of spirits would be an absolute perversion of the case, for the

red flushing of the morning was ominous, and, if I must speak the truth, it put him into that state of mind which the world have combined to call most abominably disagreeable. As he strode up Ben Derig nothing went right.—
"Davy, you are always striking the dogs with the spare ramrod. How the deuce can you be so awkward? There now, don't pull them along in that manner; they will be weary before they get half way up the mountain. Jamieson, I dare say you have left the water-proof rifle-cases at home."

"No, I have them all with me."

"Well, I did you wrong to suppose so, for I never knew you to forget any thing of consequence."

"What the de'il maks the maister so crabbed the day?"

"Crabbed, aye, and reason eneuch. The mists are rising heavier and heavier in the haugh; and though Ben Derig shimmers now, won't he be all mirk afore we can win forrat to our cast?"

And scarcely had they gained this destined point, before a great volume of mist came sailing over the lower grounds, and jostled against the huge shoulder of Ben-y-chait; then, breaking and spreading widely abroad, all around at once became dim and dubious. This was the beginning of the evil; but worse remained behind. Cloud after cloud came driving along, till the whole face of nature, mountain, rock, and glen, was smothered in the reeking vapour.

Scarcely may you discern your neighbour sitting upon the dripping heather beside you. These clouds of mist are sure to last some hours, or may continue the whole morning, and finally terminate in a deluge of everlasting rain. Sometimes, indeed, they would clear away pretty suddenly, but more often would they rise gradually. None but those who know the joys of deer stalking can tell with what an intent gaze the rifleman's eyes were fixed upon the space below him. At times the heather grew evidently clearer; then it was distinctly seen, and his hopes began to rise. The gleam was brief and delusive: again and again the huge volumes came breaking on the hill tops, and all was more sullen than ever. As for patient resignation, no sportsman knows what it means; he might possibly have read of such a thing, as Magnus Troil had of the nightingale, but certainly could not put faith in its actual existence. Once taint him with this sort of philosophy, and you ruin him for life; he is a lost man to all intents and purposes. An eager sportsman, I can understand; the phrase is apt; but who ever heard of a patient sportsman? Such a fellow would take snuff when he ought to take a snap shot; and you would see him *purgantem leniter ungues*, when he should be sweeping down a precipice like an eagle. But of such as these discourse we no farther.

Turn we now to Tortoise. Silent and abstracted he sat on the grey stone, and, passing his hand across his brows, began to brood over the scenes of his early days; again he roams over the rock-bound coast of Mull, and along the desolate shores of Iona; again he chases the roe amongst the slaty mountains and rude wildernesses of the Isle of Mist; once more he traverses the heathy Morven, and winds his solitary way amidst the rocks and hoarse cataracts of Glencoe. Here, in this birth-place of Ossian, rise up before him, in his visionary mood, the heroes of other days, the hunters of deer; and thus again he muses on that blood-stained pass :—

> Was it thy form, Fingal, that on the cloud
> Strode on as the autumnal gust blew loud,
> Deep'ning amid these rocks and glens forlorn?
> Was it the echo of thy distant horn?
> Or heard we his wild harp who drew his breath
> In the dark pass, dark as the frown of death!
> Where Cona,* creeping through the mossy stones,
> Along his gloomy way, forsaken moans,
> As if remembering still the mighty dead,
> Or mourning the fell deed that dyed his current red? †
> 'Twas not, Fingal, the winding of thy horn;
> 'Twas not thy shade wrapt in the mists of morn;
> 'Twas not, oh Ossian! thy sad minstrelsy,
> Heard o'er the mountains as the dead passed by;
> But here, as on the scene renown'd we gaze,
> Where strode the awful chiefs of other days,
> Wild fancy wakes.—Sudden before our eyes,
> As to the lonely seer that dreaming lies,
> Pale shadowy maids, and phantom chiefs, arise;

---

* A river in the pass.
† Massacre by the soldiers of William III.

> Dim floats the sombrous imagery sublime,
> Thy lone harp mingles sad its sweetest chime,
> The aged rocks seem listening to the song,
> On clouds of mist the spectre warriors throng,
> Whilst the low gale sighs, o'er their mossy bed,
> Peace to the shadows of the mighty dead!*

Break off—break off. Gone, long since gone, is that beautiful day-spring of life—alas! how fleeting—when for the time we wandered along the rude wastes and sounding shores of the stormy Hebrides, looking forward to some undefined pleasure, radiant with hope, and glowing with enthusiasm;—departed are those day-dreams of the romantic fancy;—and, the illusive veil at length drawn aside, nought is now before us but the stern realities of life.

The everlasting mist still rolls on, and although slightly ascending at times, it gives a glimpse of the dripping heather, yet another and another volume drives along, each pressing on like the waves of a troubled sea. But behold a broad white light expanding in the heavens. It is the path of the glorious sun wading in the dim expanse, and struggling with the vapour. Now it fades away, and hope dies with it:—dark—dark—dark. Oh that some blast would sweep across the moor, and scatter these lazy volumes to the four corners of the earth!

"But it will clear! I see it is clearing. Mark how the mist is gathering together, and forming in more compact masses. By heavens, it rises! How beautifully it climbs the silvery heights of Ben-y-venie! See how it courses before the sun, and how blue it is getting to the leeward!

"Shake the dew drops from your flanks, Peter; we shall start in ten minutes."

"Will you please to tak' a glass of whiskey?"

"Will I? you shall see. Out with your bottle, my good fellow: but I do you wrong, for I see it has been constantly in your hands. I only hope that it has a better smack with it than the mountain dew we have been inhaling for the last three hours. There, pass it round with wishes for

---

* The author printed these lines, such as they are, many years ago; but circumstances happened which prevented the continuance of the work in which they were included, and consequently their circulation.

success, and do not 'spill the good creature,' for in such a morning as this, believe me, it is most salubrious; manifold indeed are its virtues. What trade does it not quicken? It is a good carpenter, a good mason, a good road-maker, and a most capital deer-driver, provided it be moderately and discreetly dealt with, just as you deal with it, gentlemen."

"It is all this, sure eneuch; and I have often thought, yer honour, that the gauger who gangs intill the poor body's shieling, and taks awa his wee bit still, cannot be right at heart. It is a foul raid, and he can be no Christian. As for government whiskey, it is poor unhalesome stuff, and I wish the gaugers may stick to it; they will be sooner out of the way of honest men. But our home-made is a comfort the morn and the even, and a warm side to us o'er the moor."

"Thou art a perfect oracle, Peter; and of course thou sayest true. But it has killed many a tall fellow for all that, and taken some of your best hill companions to their last home. So now wipe my glass—no, not the whiskey glass, man, but the *prospect*, as you call it: one of Dollond's best it is, but you see there's a blear on it. And now let us start, for the glens are lit up, the sun rides high, and the day is far on. Nay, look not for deer on these heights, they will be all low down. It is useless to put off time; so forward, my lads—a good hill-man's step, long, quick, and lasting. No better way of walking when time presses. And don't be drinking out of every burn. Carry the rifles with their muzzles to the rear, and then you will not drill me with one of my own balls, as Sandy Macintosh there was near doing the other day."

"Not so near, either, for the ball didna pass within half a fut o' ye, and I didna pull the trigger,—so it wasna me that was to blame; I joost took up the gun by the neb, as she lay on the ground ahint you, and as I pulled her alang, the heather caught the wee bit trigger, and somehow or anither, she banged aff; so I couldna help it."

"Nothing can be more evident, Sandy; but only just keep the muzzle to the rear in future, and fight like the wily Parthian."

"Why, the same thing chanced to Glengarry, and he said naething ava anent it."

"Very likely, Sandy; but you see I am of a more talkative disposition; but I must tell you, that in bygone times, when a warrior came into a strange country, if he kept the point of his spear forward, he was supposed to come as an enemy, and was treated as such; but if he kept the point behind him, it was a token of friendship, and they feasted him, and gave him venison and whiskey."

"I ken that was when I was a callan, for I didn't hear aething anent it; but as the neb of the rifle is ahint, and as there is nae venison, I must tak' aff the Loch Rannoch without it."

"As in duty bound; very well, Sandy, I find thee apt."

A considerable space of ground had now been traversed without any appearance of deer, in spite of the quick and sagacious glance of the hill-man; the air had turned hot and close, and the weather was brewing up dark and heavy. Each man raised his eyes to the south and to the east, but still in silence.

"Whish—whish—down—low—I had a gliff of them in the sun blink;—hey, now the shadow is come ower: draw ahint a wee bit, we shall spy them again in the clearing;— Ou, what a dunner! They wunna bide there lang."

The clouds were now advancing in dark volumes, with their hard masses rent, as it were, from top to bottom: the thunder travelled sullenly amongst the distant chain of mountains; darker and darker still grew the huge form of Ben-y-gloe; slowly, determined, but still onward came the solemn mass; for a while it seemed to rest behind the heights of Cairn-marnoc, whilst the sun cast a last grim smile on its heathery braes.

"I am thinking we shall have a blad of weet."

"I have a slight suspicion of that myself, Maclaren, so we may as well go to Cairn Derig Beg, where the hill is steep, and we shall be more in the beild."

The rising wind came rustling on with a mournful sound; then, as it swelled into a raging blast, down at once fell the drenching torrent; and the big drops lashing along the moor, gave back a spray like the dashing of a waterfall:

louder and louder the thunder echoed from hill to hill, till it died far away on the rugged peak of Schehalien.

"I ken this Beg is no fit place for Christian men in the fire-flaught. The day is mischancy, and sure as daith something will happen, for I heard the lament sung yesterday in the gloaming, and well I ken it came from no living mouth."

"Did you see your taishe, Peter."

"I munna tell what I saw; but it was that I wudna like to see again; and sin' I hae trod the hills, I never saw sic fire as this."

The storm was indeed awful. Tortoise was sitting under the hill,—Peter Fraser was on his left,—Maclaren and Jamieson were close to his right and front, and Sandy Macintosh was with the hounds at a little distance.

The thunder clouds were now vertical; no interval between the fire and the crash, but both instantaneous, like the volleying of heavy ordnance:—another vivid flash, and a loud, piercing, and protracted shriek was heard from Fraser. The men were driven abroad, as if an engine of war had burst amongst them: each had received a violent shock—all of them in the legs; but, providentially, no one had sustained a serious injury. When the first surprise was over, they began to try their powers of moving. Fraser limped like Vulcan; but after certain moans, and a little rubbing of their legs, and skipping about to try their powers, all were soon sensible that they were as sound as ever.

It was evident, from their yelling, that the dogs had received a violent shock also.

The hurricane now bore away, raging and driving onward to the west. The peals were longer, but less loud. Then came down the rear storm in one continuous sheet of water, and soon the awful voice died away in distant murmurs. The weather gleam began faintly to appear behind Ben-y-gloe, growing more vivid as the dark mass rolled onward; at last, the sun broke forth once more—the winds were at rest—and all around looked serene and fair as in the morning. You would not have known that this thing had been, but for the small pools, or lappies, as they are called, which now glittered in the sun, and the streams working their way rapidly through the bogs, and coursing down to the burns. Those burns which but a short hour ago crept lazily through the mossy stones, were now filled with a raging, turbid torrent, rolling onwards, irresistible in its course, as the lava-streams of a volcano;—all then is passed, and the moor is still again.

"You're no thinking of the taishe now, Peter."

"Ou! but I'm thinking my legs are all arred, and that the fireflaught is in them still, and will be no be out of them the nicht; and do you no ken that yon point from which the storm came, is Cairn-na-gour, and that it was frae that vera tapmost hill that Willie Robertson, the auld forester, him that used to kill the outlying deer by Gaig, sung the lament? It was foreby that Beg he stood, and showed John Crerer the taps of a' the high hills from Aberdeenshire to Inverness-shire, and ca'd them by name, beginning at Tarff Forest in Atholl, and passing on to the taps of the Argyleshire hills, to those of Lochaber, Inverness, and Aberdeenshire, where he said he had spent mony a pleasant day. He turned round the tap of the hill, and disappeared. Crerer turned round a wee whilie after, and spied him nearly a mile aff on his way hame; he followed and owertook him, and found him sorrowful, and the tears falling from his een. He said, 'I shall never see again what I hae seen the day;' and troth, he never did. He died at the great age of ninety-two."

"Ah, poor fellow, and loath, very loath was he to leave his dear hills; for when Stewart, the ground officer, asked him if he thought himself in danger, he said that he knew he was dying, and that he had little chance of ever seeing the Duke again in this world; but he hoped that when his Grace was taken away also, he would meet him at heaven's gate, and welcome him in. He then began praying; and, in the middle of his prayer, asked Stewart, 'if it was true that his Grace was going to make a road up Glen Mark and Glen Dirrie.' Stewart told him 'that was only a joke.' William answered, 'that making the road would be no joke.'

"But he enjoyed a long and happy life, and I hope you

will not sing your coronach at an earlier age. It is a custom, I believe, which all the old foresters have observed. I was near hearing poor gallant Macintyre sing his: you may remember when he was lying ill at Forest Lodge, and I had my quarters there, how, in the midst of his fever, he would rave about the deer; how his spirit was ever on the hills, whilst his body was lying on a sick bed; how wildly he talked of Ben-y-gloe, Craig-chrochie, Glen Croinie, and all the glens and mountains that had so often echoed to the crack of his rifle; you may bear in mind how near he then was to the grave of his fathers. It chanced I did him some little common act of kindness, such as no one but an honest-hearted Highlander would have thought about for a moment. He wished, he said, he might get well, that he might have the pleasure of taking me into the deer—how fine he would do it! These were the last words I ever heard from his mouth, and surely they were kind ones. Poor fellow! on that day I sent him down to Blair, in an easy carriage, to be nearer the doctor: he lived but a short space afterwards. Long before this, however, he was aware that his life was ebbing; for when Mr. Landseer painted his portrait, he looked at it sorrowfully, and said, 'An' if that's like Macintyre, he's no long for this world.' Too truly did he prophesy,—peace be with him.

"And now we will see if we can kill a hart in honour of his memory; and we will pour over the best libations of right Loch Rannoch, the fumes whereof will be grateful to his shade."

Peter Fraser (touching his cap), "That would be shamefu' waste, yer honour; Macintyre himsel' aye poured it intill his weem, and I'm thinking his ghaist would like to see us pit it in the same gait, and not gie it to a dead beastie, who will no ken whether it be lowland stuff or richt Loch Rannoch." *(Then laying his arm upon Tortoise),* "Hist, hist, sir; some fashious body has disturbit the moor. Look to yon deer; they are coming ower from the east by the green knowes, and ganging on slowly to Crag Urrard. What shall we do? We maun lie doun on the heather, for we are lockit in, and canna win forrat a fut the noo. The banks of the Banavie are steep, and the pass to Crag Urrard is narrow;

but we are ower far awa' to rin intill them at ony gait;
but your honour gangs wi' lang strides doun the brae, and
ye may mak a push for it when they are ower the hill; but
ye maun gang your best."

"They are going slowly, Peter, and I do not altogether
despair; it is a long run, but we have no other chance at
any rate. The worst of it is, that this long heather, which
appears so even, is full of large grey stones, that lie hid in
it on purpose to break honest hillmen's legs, and yours are
all arred with the fire-flaught, you know, Peter. But we
will not heed a sprained ankle or broken leg or two in such
a cause, though the chance be a wild one."

Tortoise now began to measure with his eye the long
distance to the pass, which seemed to be about a mile and
a-half, and then to consider how long the deer would probably be in crossing, after they had sunk down the hill out
of sight of the stalkers; it would be a race against time,
and his calculation was an unfavourable one.

In the midst of this anxiety they had not observed that
the weather was again brewing up in the south; and the
rain began to fall thick and heavy: they now judged that
the deer had not been disturbed by any traveller, but were
slowly shifting their ground to get under the hill to the
leeward, for they did not look back to the point from which
they came, or show any jealousy; neither were they in any
hurry, but walked slowly, stopping occasionally to feed.
During this tedious time the rain fell heavily, and came
trickling through the bonnets of the recumbents. Could
they have been posted in concealment one short half-mile
nearer, all this they would have borne patiently, as they
had borne it many a time and oft. But now that the
chance was almost nothing,—cold, rheumatism, and all the
ills that flesh is heir to, appeared in sad and hideous array
before Tortoise's imagination; and, as the cast was now
nearly ended, the base thought of going homeward, without
waiting for the chance, came across his mind.

Hear it not, O noble shade of stout Glengarry; you who
would lie abroad in cavern or in moss for nights together,
the grey stone or the drifted snow your pillow; you who
would swim through lakes and flooded rivers, alike heedless

of the tempest and all the barriers that rugged nature threw across your course—hear not, I beseech you, the recreant thought that came across our mind. Alas, had not your generous spirit departed prematurely; had not the mournful sound of your coronach been borne on the hollow blast through your rocky glens and mountains, lamented as you were by many a true heart and brave clansman; oh, had you still lived, buoyant in all your strength and national spirit, I would have sung *Io Pæans* to your triumphs; though, after this confession, candid as it is, your heart never would have warmed towards me again, which you once told me it did, as being the descendant of a borderer.

The thought of going to the halls of Blair, however, with the deer in view, was transient as it was degenerate; and, to do ourselves justice, never would have occurred to us for a moment, had not the cast been nearly finished, the chance almost as nothing, and had not visions of warm fires, hospitality, and happiness floated invitingly around the old towers of the castle, already in view; in other words, had we not at that moment been somewhat of a milksop. A blush came over our storm-beaten cheeks; we vamped up our wet courage, and were determined to await the event.

Long, very long, did the party remain under the wrath of Jupiter Pluvius; for the pasture was good, and the deer were in no hurry to quit it; and, as the men were locked in, they could not move till the deer did. At length they began to draw on slowly over the hill; two or three disappeared, others followed, but more lagged behind.

"Will they never go? Yes, yes, they seem to be all drawing on; and now, by Jove, they are all fairly over, except that jealous hind. Fix your glass steadily upon her, Peter, and do not speak till she shall be clearly out of sight when we are standing up; try it first on your knees, and raise yourself slowly."

Fraser looked awhile, then shut up his glass rapidly, saying, "Noo's yer time, she is clean awa'."

Up they sprung, and away they went at high speed, steeped and drenched as they were with the rain, which had never ceased for a moment. Sometimes they stuck a'most knee deep in old heather, amidst large blocks of stone: these they sprang upon, or twisted their ankles between, as it happened; for such a swinging pace down hill precluded their arresting their steps for a moment. Soon they come to the great declivity, and look anxiously to the opposite steeps of the Banavie. The deer were not going up: they had them below then; but the descent was long, and they might still be baulked. Down—down they rush behind a ridge of ground, stooping and peeping just to the north of the spot where the deer had passed.—" And now they cannot escape us; we must have them, for good or for evil."

The tops of their antlers were just in sight; and down dropped the men at once, motionless, in the heather.

The deer now advanced through the burn: Tortoise singled out the best harts, keeping his eye steadily upon them, and marking their precise course; but, as yet, moved not his rifle. They dashed, and splashed, and shifted places in such a manner, that he judged it most prudent to wait till they were ascending the steep. He then had nearly their whole backs presented to his aim. When they were in this position, up at once he sprang, and discharged his three rifles in succession.

At the first shot, a magnificent hart sunk down upon his hind quarters, staggered, and rolled back lifeless into the burn: the ball of the second rifle passed down through the shoulders of another splendid fellow; he fell forward, and was instantly dead. The last shot was fired too high, and only cracked against a stag's horn, which stunned him for a moment, but he soon recovered, and went off with the rest as well as ever.

Nothing could surpass the joy of the party at this almost unhoped-for success: they canvassed the thing over and over again. It was a wonderful distance to come in from; they never ran so well in all their lives;—in short, they were prodigiously fond of themselves,—especially as they had anticipated a blank day. They never chose to consider that the deer (who had not seen them till the last) were going very leisurely.

"Out now with the whiskey bottle, man, and we will

make our promised libations in favour of the good old forester."

Whilst honest **Davy** was extracting this desideratum from his pocket, one of the dogs slipped his collar, and seized the throat of the hart, which the men were lifting out from the burn, with savage ferocity; being choked off when they gained the banks, he turned his wrath towards his friend in the leash, and these two bloodthirsty villains flew furiously at each other, and were parted at some risk and difficulty. This sort of conflict was, indeed, a very common occurrence; it began with a low growl, then a grinning, and exposition of certain white teeth; then a setting up of bristles, a sudden spring, and war to the knife.

"Now, then, all hands to work, and let us see if the fat of this fine fellow is bruised by the fall. No, I am sure it is not; he feels quite firm and sound. Davy, you rogue, put the quaigh in your pocket, and gralloch the other deer, whilst we attend to this."

The harts fell near the pine woods of Blair: a smart walk, varied with an occasional run, put in practice when their late feat came vividly over them, soon brought them to Blair. They no longer heeded the rain and the blast, but now rather rejoiced in it.

"Forsan et hæc olim meminisse juvabit."

CHAPTER VI.

The Forest of Atholl.—Probable number of deer, and their size.—Cumyn's cairn.—Highland vengeance.—Fatal accident.—Principal glens.—Glen Tilt.—Marble quarries.—Roe deer.—Lakes and lodges.—Merry foresters.—Forest song.—Cuirn-Marnich.—Last execution at Blair.—Arrest of a murderer.—Royal feasting and hunting.—Palace in the forest, and Highland cheer.—Burning of the palace.—Kilmavonaig beer.—Cumyn's death.—Belief in witchcraft.—M. G. Lewis's legendary tale of the Witch of Ben-y-gloe.

"There's the dae, **the rae**, the hart, the hynde,
 And of a' wild beastis great plentie;
There's a fair castell of lyme, and stane,
 O gif it stands not pleasauntlie!"

Minstrelsy of the Border.

The celebrated forest of Atholl comprehends a vast tract of moor and mountain, extending, by hillman's computation,

from the north-east point joining Aberdeenshire, to the south-west point joining Gaig Forest, about forty miles in length. The extreme breadth, from the top of Skarsach, north side of Tarff, to Craig Urrard, Mr. John Crerer thinks cannot be less than 18 miles, but it diminishes in breadth at the extremities. It measures 135,451 imperial acres.

The following table will show how it is divided and appropriated:—

Contents of the Atholl Deer Forest, &c.

| | Imperial Measure. ||
	Forest Ground. Acres.	Grouse Ground. Acres.
1. Glenfernate	-	10,720·15387
2. Felaar and Tarff	10,089·46760	11,350·65105
3. Glentilt, Benygloe, and Loch Valligan	26,484·85245	5,044·73380
4. Riechlachrie and Benychatt	10,089·46760	-
5. Glenbruar	5,044·73380	7,567·10070
6. Glen-Crombie and Kyrachan	-	7,567·10070
7. Aldvoulin and Cluns	-	8,828·28415
8. Dalnacardoch and Wood Sheal	-	6,936·50897
9. Dalnaspidal and Mealnaletroch	-	18,287·16000
10. Bohespick and Strathtummel	-	7,440·98235
Total	51,708·52145	83,742·58959

Total { Forest Ground 51,708·521 / Grouse Ground 83,742·589 } 135,451·110.*

The part of the forest which is kept for deer-stalking, it will be remarked, is 51,708 imperial acres, and is bounded chiefly on the west by Craig Urrard and the river Bruar; on the north by the Tarff; on the east by the Felaar grouse ground; and on the south by the cultivated grounds and woods of Blair. Deer, however, are occasionally to be found beyond these limits—particularly hinds: in a north wind, indeed, an inexpert or rash deer-stalker would send vast numbers out of this ground, and if they were still pursued and followed with pertinacity, they might be driven into other forests and remain there some time.

---

* The names of the various hill tops are given in the Appendix.

All this vast tract is reserved exclusively for deer, with a slight exception as to Glen Tilt, where sheep are occasionally permitted to pasture. In 1786, the sheep were removed from the north side of Glen Tilt, and from the south, or Ben-y-gloe side, about ten years afterwards. In the year 1776, when Mr. John Crerer went to Blair, the number of deer in all the forest did not probably exceed 100; though some small herds have wandered in it from time immemorial. The great increase took place in the year above mentioned, when Forest Lodge was built; the sheep and cattle were removed, and the hills were thus kept free from disturbance. Favoured and protected as they now were, the increase became very rapid; so that of late years their numbers were computed at about seven thousand; but I always thought this an exaggerated statement; for I once saw on the same day all the deer driven down from the east, and a second drive also from Glen Crinie; I then fell back north before the deer had crossed Glen Tilt, and came to Blair by the western cast and the lower grounds; so that, with the exception of such as happened to be on Ben-y-gloe, I must have then seen almost all that the forest comprehended, as the wind was full south; making all allowances, I should estimate the number at between five and six thousand. On this day I killed seven fine harts. The hinds are of course far more numerous than the harts, as none but yeld hinds are killed, except by accident. It must be allowed, however, that these accidents happen pretty often, and indeed, in almost every deer-drive; for young sportsmen will fire at all hazards when they have rifles in their hands—aye, and old ones too, sometimes.

It is thought that the harts in Atholl Forest are inferior in point of size to those in other districts; and from the weight of stags killed elsewhere, an account of which has been sent to me, I am forced to come to the same conclusion. As the pastures are excellent almost everywhere, and particularly rich on the north brae of Ben-y-gloe, this inferiority in point of size cannot be attributed to the incapacity of the ground to produce larger animals. It arises, I think, from a very obvious cause: Blair being in the high road to the north, almost every sportsman that came from

England profited of its hospitality, and participated in its amusements; thus there never was a day in the season when the wind was favourable, in which the deer were not disturbed to the utmost limit that the forest would admit of. Some of the best harts were killed off, to the number of 100 or 130, or perhaps more, in each season; and many others, I imagine (and these the largest), found their position so unquiet, that they sought the forests of Gaig and Braemar, and deserted that of Atholl, where they were continually driven, and kept in a state of perpetual alarm. It is evident that no animal could arrive at his proper dimensions under such harassing circumstances.

But many people were deceived as to the actual size of the Atholl harts, from the custom of reckoning by Dutch weight, whilst others used the imperial. Now as Dutch weight is seventeen ounces and a half to the pound, and sixteen pounds to the stone, the difference is most material. The weight, too, was given not as the deer stood, but after he had been gralloched.

But if the pastures are fine, the ground also is in all other respects the most favourable that can be imagined for a forest. Mountains of various altitude, open sunny corries, deep glens and ravines, holes for solitary harts to hide in, and numerous rolling pools, burns also and rivers, and large pine woods to shelter them during the inclement season.

The two highest mountains in the forest are Ben-y-gloe and Ben Dairg, or the Red Mountain. Ben-y-gloe is of vast magnitude, and comprehends a little territory within itself, stretching its huge limbs far and wide. It is computed to be twenty-four Scotch miles in circumference, and it contains twenty-four corries; these corries are separated from each other by such high ridges, that a person standing in one of them could not hear a shot fired in the next. The highest point of the mountain is Cairn-na-gowr, or the Goat's Hill, which is 3725 feet above the level of the sea. On the eastern side of Ben-y-gloe lies Loch Loch, abounding in char and trout; and near it stands Cumyn's Cairn, concerning which tradition has given us the following story:—

About the beginning of the thirteenth century the authority of the district was divided between the family of

Cumyn Earl of Badenoch, and M'Intosh of Tirinie. The latter had presented Cumyn's lady with a present of twelve cows and a bull; but this substantial donation, so far from exciting the gratitude of the chief, only raised his envy and cupidity, and he resolved to strip his neighbour of his opulence. He surrounded M'Intosh's castle of Tomafour, situated about a mile from his own castle of Blair Atholl, and in the silence of midnight massacred the whole family. Near M'Intosh's seat lived an old man who held a piece of land of him, for which he only paid the rent of a bonnet yearly; and he always got his master's bonnet back again. This man was the first who entered the castle after the murder, and casting his eyes round on the scene of death, fortunately discovered an infant sleeping in its cradle. He carried away the child to its nearest relative, Campbell of Achnabreck, in Argyleshire, and there the boy was nurtured, unconscious of the melancholy story of his parents. It was judged prudent to conceal his birth for some time, as the Cumyns were a powerful race, whom it was perilous to offend. The boy grew, and became an excellent bowman; his aged conductor used to go occasionally to see him, and perceiving his dexterity in hitting the mark, said one day, 'The grey breast of the man who killed your father is broader than that target.' This led to a recital of the whole transaction. Even the young laird burned for revenge; and he succeeded in obtaining a select band of clansmen to share in his feelings. They went to Cumyn's castle, and assailed him with a shower of arrows. His followers were scattered, and the guilty chief fled to Loch Rannoch, Glen Firnat, and thence to Glen Tilt, hotly pursued by his much injured adversary. At length, as he raised his hand to wipe the sweat from his forehead, he was struck with an arrow, and fell by the side of a small lake at the foot of Ben-y-gloe, where a cairn was raised to perpetuate his crime and its punishment.

The above story is yet current in the country, and the remains of M'Intosh's castle may still be seen. There is a rock in the Tilt called M'Intosh's Chair, where he held his court, his people standing round him; happily he could only do so when the water was very low, as he hung a man every time. It is still a bye-word in the country that "It is not every day M'Intosh can hold his court."

A fatal accident happened at Craigantsuidh, near Poll Tarf, about sixty or seventy years ago; and here follow the particulars, as I have received them:—

Alexander Macgregor, a resident in Glen Tilt, was travelling with two companions on the face of Craigantsuidh, which is very rugged and precipitous. It was at that time covered with snow, and sheets of ice were found in various places, which frequently conducted to the ledge of a precipice. In an evil hour Macgregor, unconscious of the danger, placed his foot on one of these perilous spots, which conveyed his body over a deep precipice, and his soul to eternity.

His two companions took his corpse into a shepherd's hut, where they proposed leaving it that night, until they procured assistance.

The one said to the other, "Will you go to Felaar for assistance, or remain all night with the body?" He replied he would go to Felaar for assistance. The Camerons were there at this time in the capacity of foresters. He was scarcely gone, ere the man, who remained with the body, was pelted with stones and turf, and other missiles, till he was provoked to go out and see from what direction they were thrown. On his going out they ceased; but the moment he re-entered, they began again with such increased violence that he would have been stoned to death had he not left the house.

The country people attributed this attack to the omission of leaving the door of the hut open to give a free passage to the departing spirit. People will form their own conclusions on this and similar stories. I mention them as evidence of the superstitious feeling that still pervades some secluded spots in the north.

In the year 1804, one Duncan Robinson had a narrow escape from the fall of an avalanche on Ben-y-gloe, but (more fortunate than Macgregor) he saw the impending mass of snow tottering above him, and threw himself under a rock that was providentially by his side; the vast volume passed over him, and his life was thus spared; but his dog perished.

The principal glens in that part of the forest, which is set apart for deer, are the celebrated Glen Tilt, Glen Croinie,* Glen Mark, Glen Dirie, and Glen Bruar,—all bearing the names of the rivers that run through them; and all which rivers (save the Tilt) run from north to south nearly in a parallel direction.

The Dirie falls into the Mark; the Mark rushes into the Tilt; and the waters of the Bruar lose themselves in those of the Garry. The Garry itself may also be considered as within the precincts of the forest. Towards the north is the Tarff, which runs nearly from west to east with a bearing towards the south; and it falls into the Tilt at the head of the glen. The Croinie also falls into the same river. There is likewise a lesser stream, called "Auld Banavie," on the western side of the forest, which runs beneath Craig Urrard; the latter part of its course is full of wild and picturesque character: it is swallowed up in the waters of the Garry.

Salmon come up the Tilt in full waters, and are taken with the fly; and all the other rivers are so full of small trout, that any one who pleases may catch as many dozens in a day as he can conveniently carry. These streams work their way in solitude through dreary mosses, and come winding down the glens sometimes in comparative tranquillity, and at others bursting and rooting up every thing about them; the mighty force with which they descend may be read in the vast rocks and fragments of wreck which they heap up as monuments of their power.

Supplied by such numerous forces, the Tilt becomes powerful in its infancy. Born in rugged regions, it cleaves its way, at the base of impending mountains and rocky precipices, in a dark, deep, and narrow trench. Arrived at the green pastures of Ben-y-gloe, its bed begins to expand, and the waters pass down in a freer course; still however they come racing and flashing along with overwhelming violence.

A little lower its wrath is tempered with all the ornament that art and nature can bestow. First of all a few straggling trees deck its margin; then groups of birch

---

* This is usually pronounced Glen *Crinie*.

stand airy and light, displaying their glossy stems upon the knolls, or shelving down the sides of the great mountain, vivid as it here is with luxuriant pasture. The woods now skirt the braes in larger masses, winding on the hill sides and conforming themselves to the varied undulations of the surface. They press closely on the river where the valley is contracted, and their branches wave over it, and shed the sear leaf in the stream. Some of the masses are dense; others admit the sunbeam, striking on the scarlet berries of the mountain ash, and bringing out the rich autumnal tint of the brachen which grows beneath them. All soon uniting in mass, gather into larch and pine forest, and at length mingle with the woods of Blair.

The pass itself is barred in by the grim mountains that heave their dun backs about it, and send down many a torrent from their riven ribs. A good road winds along the braes, catching and losing the waters as it pierces the gloom of the woods, or breaks forth into light and expanse. Picturesque bridges are thrown across the river, and every thing has been done that consummate judgment could effect to temper the wild scene with beauty and convenience; to temper, but not to destroy it; that indeed, if advisable, were almost beyond the power of man. Stern and indomitable as the wrath of Achilles, the Tilt ever holds its mood, and comes raging on, wheeling in eddies, rushing in cataracts, or spreading into pools, bearing along with it at times huge fragments of rock that form uncouth islands in its channel, upon which the stricken deer stands dominant at bay; still ceaseless it races onward, fretting and foaming, till at length its mad career is arrested in the less turbulent waters of the Garry.

After the storm this river speaks in a voice of thunder, and quells every noise around it; but when the winds are hushed, and the weather gleam streaks the sky from afar, and the rain-drops glitter in the sunshine, some sylvan sounds may occasionally be heard—the solitary croak of the raven's voice as he sits boding on the crags, the distant bellow of the hart, or the scream of the eagle falling faintly on the ear from the skies above. In a grey day the mountains around are stern and dark, and there is gloom all up the glen; so that when the eye travels to the small opening at the distant gorge, you look out at the bright light of heaven as from the mouth of a cavern.

But it is in the clear day of autumn that this scene is most enjoyable, when the air is invigorating, and when the sunbeams strike down the summits, and the light falls on the glossy stems of the birchen grove, warms the grey rock and the greensward, and brings forth all the rich hues of decaying foliage. Yet even in the broad evidence of a meridian sun, whilst the light leaves tremble and sparkle in its beams, and countless objects stand prominent, luminous, and defined, there are vast masses of dark pines unrevealed and impervious to its genial influence, and deep flat shadows that leave much in mystery and obscurity.

The whole of this glen, in a scientific point of view, is interesting in the highest degree; to a geologist there is none more so throughout Scotland. A quarry has been opened above Marble Lodge, which contains immense blocks of marble, varying from grass green into one of a yellower cast, and intermixed with grey. The best blocks take a good polish; and it surpasses in beauty all analogous subjects of British origin. The transportation of such a heavy material however is not easy, as the Tay is not navigable above Perth.

There is also a beautiful yellow marble to be obtained, which is mottled with white; as likewise a coarse sort of white marble polluted with grey streaks.

In the forest there are four mineral springs: I am not aware that they have been analysed, but many an incautious hill-man can attest their efficacy. One of them issues at the side of Loch Mark, one at Dualdan, north from Felaar House, and two at the top of the burn, at Inverslanie. The two last are named Duke James's Wells.

There are a great many roe deer in the forest, which feed chiefly in the woods, or on the moor immediately adjoining them, but are never seen far out on the hills. They do not unite in herds, but live in separate families. In favourable seasons, about one roe out of five or six will produce two fawns.[*] As a singular proof of their attachment to

---

[*] Various writers make the proportion of twins much greater, but this is Mr. John Crerer's calculation.

their young, I here transcribe an occurrence that has been obligingly sent me by my eminent friend, Sir David Brewster.

"Near Belleville, in Inverness-shire, there is a finely wooded range of rocks, containing Borlam's* Cave; the haunt of the last Highland cateran, who was proprietor of Belleville. In cutting a path to this cave, one of the party of Highland labourers, whom I took with me for that purpose, asked me if I had seen the spaning (weaning) tree of the roe deer, and pointed out one close by us, which, but for this notification, would have fallen under the axe. This tree was a small birch one, that stood nearly in the middle of a regular oval ring, formed and trodden down with the feet of the roe deer, who run round and round the tree, followed by their young, in order to amuse them at the time when they are weaned. My informant assured me that he had seen the deer engaged in this sport, and I have myself seen and shown to others the footmarks of the old and young deer in different parts of the ring round the birch tree; at one end of the ring there was a small oval, giving the whole the appearance of the figure ∞.

These beautiful animals, however, who for the most part lead such a tranquil and domestic life, are animated with fury like the red deer during the season of rutting. In the summer of 1820, two roebucks were discovered in a deep hollow, one above the other, most firmly united in the following singular manner:—The horns of the uppermost one were twisted in the skin behind the shoulders of the one beneath, and those of the latter were twisted in a similar manner in the shoulders of the buck above him. Both were found dead in this dreadful position.

There are seven lakes in the forest—Loch Tilt, Loch Mark, Loch Garry, Loch Hone, Loch Dhu, Loch Maligan, and Loch Loch; the last abounds in char, and on its banks stands Cumyn's Cairn.

---

* Tradition says, that, whilst this ruthless villain was in the act of burying a man whom he had robbed and murdered, he was discovered by a clansman, who rebuked him. Afraid of legal retribution, he struck the intruder down with his spade, jammed him at once into the earth, and buried both bodies in the same grave.

There are two hunting lodges in Glen Tilt—Forest Lodge, and Marble Lodge. The latter is a mere station; the former was built in 1776, and has lately been much enlarged. It is constructed without affectation of ornament, and consists of two tenements united by a stone screen surmounted by stag's horns, and in which there is an archway for carriages to pass. One of these tenements serves for the lord of the forest and his friends, and the other for his retinue. The foresters and gillies, however, are so numerous, that I have often wondered by what means so many human beings could be packed together in so close a space. So it is, however: and instead of complaining of inconvenience, every man is as happy as if he were sole possessor of the great bed at Ware. As a proof of this good feeling, and the general spirit that pervades the hill-men, I transcribe a song made by Alexander, an old and faithful servant of the late Duke of Atholl, who lived with him eighteen years, and now lies buried in the cathedral at Dunkeld. This composition was sung every night at Forest Lodge when Maddy was there; and, whatever may be thought of the poetry, is as good an evidence of the sort of thing going on as I can possibly give. Here it is in its pure doggerel state. I have not attempted to spoil its character by the alteration of a single word :—

ON SEEING LORD CATHCART ARRIVE AT BLAIR ONE MORNING EARLY.

O, Campbell*, man, I muckle dread
  That we shall have a tramp;
The Commander in Chief † so soon a stear,
  I fear we must flit our camp.

But if to Felaar we do march off,
  As I muckle dread we may;
Some Athole brose before we go
  Campbell and I shall ha'e.

---

* Campbell was cook in the Atholl family upwards of sixty years; but for several years before he died acted as hill cook only.
† Lord Cathcart was commander of the forces in Scotland at that time.

The journey's long and rugged too,
　　Some waters for to cross ;
Some hills to climb—but worst **of all**—
　　Is troughing through the moss.

When at Felaar we do arrive,
　　How pleasing 'tis to see
At night the harts and birds come home,
　　In dozens **twa or** three.

John Crerer he spies out **the harts,**
　　My Lord Duke does shoot them ;
Curly* he does bring them home,
　　And Campbell he does cook them.

Tho' **Campbell** carries nothing there
　　But just a pan and brander,
He can soon cook a dinner rare
　　For the **Duke** or Alexander.

And when our kites is a' **weel cramm'd**
　　Wi' ilka thing that's **rare,**
Then to the toddy **we sit doun**
　　Each man to drink his share.

Lang life to you Campbell,
　　To stear about the toddy ;
Of a' the friends I ever ken't
　　Ye are a dainty body.

Next to bed we do prepare
　　The best way we are able ;
There is twenty **lies** upon the floor,
　　And Maddy on the table.

From wa' to wa', all in a row
　　Like **herring on a** plate ;
The man that durst our camp attack
　　My faith he'll **no** be blate.

Such a regiment of Highland men,
　　The Duke and Lord Cathcart ;
I am convinced they would defy
　　The devil and Buonaparte.

---

* John Forbes, christened "Curly" by the Duke, from his hair being much curled, attended his Grace upwards of twenty years to the hill with two horses, to bring home the dead deer to Blair. This man knew every part of the forest, and could be directed to find the dead deer, though lying twenty miles distant from Blair. He died about the year 1825, aged about seventy.

Ben Dairg, or Derig, as it is usually pronounced, the mountain next in consequence to Ben-y-gloe, is 3,550 feet in height. It lies about ten miles north from Blair; its summit is covered with immense blocks of gneiss and granite of a reddish colour, from whence it derives its name of the Red Mountain. This chaos of huge fragments is the favourite haunt of the ptarmigan and white hare, though the perilous den of the fox and wild cat is there also, and the eagle preys around it.

The south side of this mountain forms a vast crescent, the horns lying west and east.

I must not omit to mention more particularly another mountain which lies between Glen Mark and Glen Croinie: it is called Cùirn-Marnich; cuirn is the plural of cairn, and marnich of maronach—"the cairns of the Braemar people." These cairns are sixteen in number, and were raised by the Atholl men to commemorate a victory they obtained over the Braemar people, whom they here overtook and slew to the number of sixteen, as they were returning home with plunder from their country. Tradition says little about this foray, which, indeed, was but upon a small scale. It is a boast of the men of Atholl, that they never were beaten by their neighbours in open fight, such having always proved fatal to their adversaries; so that the only loss they ever suffered was by stealth and stratagem.

This they are still proud of. Alexander Gon, from Blair, was once in Braemar, when the company he was with began to banter the Atholl men for lack of courage. Up he started on his legs, and striking the table with his clenched knuckles, exclaimed, in the stern spirit of a clansman, "Remember, lads, who have the Cuirn-Marnich." This effectually silenced the banterers.

Turning from such lawless proceedings, I will now give an account of the last public judicial execution that took place at Blair.

About the beginning of the sixteenth century, the Earl of Atholl had two foresters named Stewart and Macintosh; the former resided at Auchgoul, and the latter at Dalnachie, both in Glen Tilt. Macintosh had also a bothy at Coirrerenich on Ben-y-gloe, where he occasionally slept.

One day after shooting together, they resolved to sleep at this bothy. They had only a gilly, or servant, with them belonging to Macintosh. The two foresters slept in one bed, and the gilly in another. Whilst Macintosh slept, Stewart stabbed him with his skiandubh; and, going to the gilly's bed, stabbed him also, and put him on Macintosh, that it might be supposed one had killed the other. He left them both for dead, and made the best of his way home.

Soon after his departure, the gilly began to recover a little from the wound he had received, and contrived to crawl on his hands and knees to Dalnachie, which he reached next morning, and gave information of the murder of his master. When Stewart heard that the gilly had escaped with life, and that the murder was discovered, he fled to Lord Reay's country in Sutherland, which had the privilege in those days of protecting criminals from justice.

In the meantime the Earl of Atholl, being informed of what had had taken place, was determined to bring Stewart to justice, and sent a man named Macadie, who knew Stewart well, dressed as a beggar, to discover if he was still in Sutherland. He soon returned with intelligence of Stewart's being there, and the Earl sent a strong party with Macadie (still disguised as a beggar) with instructions to bring back the murderer dead or alive.

On the arrival of the party at Lord Reay's country, hearing that Stewart was to attend a wedding on a certain day, they agreed to surround the house where the ceremony was to take place, pretty late at night. Macadie was to enter and drink Stewart's health; and this was to be the signal that the person they sought for was within. This being settled, Macadie entered, and drank Stewart's health, who, finding he was discovered, bolted out of the house, and was immediately surrounded and secured by the party stationed without, who took him to Logierait, where he was confined some time, and finally condemned to be hung and gibbeted. The sentence was carried into execution at Blair; and this man was the last person who was hung there.

The motive for this foul act did not transpire; but it was supposed that it was perpetrated for the sake of involving the murderer with the sort of horrid consequence that was attached in those days to the most daring delinquents. An obelisk was placed on the spot where the execution took place, by Duke James, in 1735; and the mound is still called "The Hangman's Mount."

The forest of Atholl seems to have been celebrated for the sports it afforded for many ages.

King Malcolm, called **Cean-Mohr** (great head), who reigned in Scotland from 1056 to 1093, frequently hunted in it; and many places in the forest are named after him, such as the King's Cairn, etc.

The Lord of Atholl Forest has the privilege of hunting over the Lude property; and the proprietor of the latter is obliged to keep his ground clear of cattle and sheep for the space of three weeks previous to a grand hunt, if desired to do so. This right was kept up for a considerable period, but has not been exercised of late years.

In Piscottie there is a description of an entertainment given to royalty by the third Earl of Atholl, which, however well known, is of so splendid and unusual a character, and so directly to the purpose, that I cannot, I think, omit it with propriety.

"In 1529, King James the Fifth passed to the Highlands to hunt in Athole, and took with him his mother Margaret, Queen of Scotland, and an ambassador of the pope, who was in Scotland for the time. The Earl of Athole hearing of the king's coming, made great provision for him in all things pertaining to a prince; that he was well served and eased with all things necessary to his estate, as he had been in his own palace of Edinburgh. For I heard say this noble earl gart make a curious palace to the king, to his mother, and to the ambassador, where they were so honourably eased and lodged, as they had been in England, France, Italy, or Spain, concerning the time, and equivalent for their hunting and pastime; which was builded in the midst of a fair meadow, a fair palace of green timber, wind (1) with green birks (2) that were green both under and above, which was fashioned in four quarters, and in every quarter

---

1 "Wind," Wound, or bound.
2 "Birks," Birch trees.

and nutre thereof a great round, as it had been a blockhouse which was lofted and geisted the space of three house-heights (3); the floors laid with green scharets (4) and spreats (5), medwarts (6) and flowers, that no man knew whereon he zeid (7) but as he had been in a garden. Further, there were two great rounds in ilk side of the gate, and a great portcullies of tree falling down with the manner of a barrace (8), with a draw-bridge, and a great stank of water of sixteen feet deep, and thirty feet of breadth: and also the palace within was hung with fine tapestry, and arrasses of silk, and lighted with fine glass windows in all airths (9); that this palace was as pleasantly decored with all necessaries pertaining to a prince, as it had been in his own palace-royal at home. Further, this earl gart make such provision for the king, and his mother, and the ambassador, that they had all manner of meats, drinks, and delicates, that were to be gotten at that time in all Scotland, either in burgh or land; that is to say, all kind of drink, as ale, beer, wine, both white and claret, malvasy (10), muskadel, hippocras, and aquavitæ. Further, there was of meats, wheat bread, mainbread, and gingerbread; with fleshes, beef, mutton, lamb, veal, venison, goose, grice (11), capon, coney, cran, swan, partridge, plover, duck, drake, brisset cock (12), and pawnies (13), black cock, muirfowl, and capercailies. And also the stanks that were round about the palace were full of all delicate fish, as salmonds, trouts, pearches, pikes, eels, and all other kinds of delicate fish that could be gotten in fresh water, and all ready for the banquet; syne were there proper stewards, cunning baxters (14), excellent cooks and potingars (15), with con-

---

3 "Three house-height," **Three storeys high.**
4 "Scharets," Green turfs.
5 "Spreats," Rushes.
6 "Medwarts," Meadow-sweet.
7 "Zeid," Sat.
8 "Barrace," Barrier, an outwork at the gate of a castle.
9 "Airths," Quarter of the heaven; point of the compass.
10 "Malvasy," Malmsey wine.
11 "Grice, or Gyree," A young wild **boar.**
12 "Brisset Cock," Turkey.
13 "Pawnies," Peacock.
14 "Baxters," Bakers.
15 "Potingars," Cooks who **prepared herbs.**

fections and drugs for their desert. And the halls and chambers were prepared with costly bedding, vessels, and napry, according for a king; so that he wanted none of his orders more than he had been at home in his own palace. The king remained in this wilderness at the hunting the space of three days and three nights, and his company, as I have shown. I heard men say it cost the Earl of Athole every day in expenses a thousand pounds.

"The ambassador of the pope, seeing this great banquet and triumph which was made in a wilderness where there was no town near by twenty miles, thought it a great marvel that such a thing could be in Scotland, considering how bleak and barren it was thought by other countries, and that there should be such honesty and policy in it, and especially in the Highland where there was but wood and wilderness. But most of all, this ambassador marvelled to see when the king departed, and all his men took their leave, the Highlandmen set all this fair place on a fire, that the king and the ambassador might see it.

"Then the ambassador said to the king, 'I marvel, sir, that you should thole (16) yon fair place to be burnt that your Grace hath been so well lodged in!' Then the king answered the ambassador, and said, 'It is the use of our Highlandmen, though they be never so well lodged, to burn their lodgings when they depart.' This being done, the king returned to Dunkeld that night. I heard say that the king, at that time in the bounds of Athole and Stratherne, slew thirty score of harts and hynd, with other small beasts, as roe and roebuck, wolf and fox, and wild cats."

In the description of the Badenoch country I have recounted a story of Walter Cumming, who was killed by a fall from his horse the day previous to an infamous exhibition which he meditated. The story is given precisely according to the belief of that district. I have since received more particulars of that event from the Atholl country, and from a source wholly unconnected with the previous one. The Badenoch authority says that Cumming was absent on some business in Atholl.

---

16 "Thole," To bear with, not to oppose.

The tradition is that he was attempting to make a road between Blair, Atholl, and Badenoch. And the cause of his undertaking so enterprising a work is thus given, though probably his real reason was of a predatory nature :—

Cumming and his wife (who were from Ruthven, or Ruairm, in Badenoch) were passing through Atholl, and on their arrival at Kilmavonaig, they went to a public-house to take some refreshment. On their entrance they called for some beer, which was then the chief drink of the Highlands; and being exceedingly pleased with it, were anxious to know where the several ingredients could be procured. The landlord, who, like Boniface, was loud in commendation of his own beer, told them he received the malt from Perth, and the water from Aldnehearlain (a small rivulet which runs through Kilmavonaig), which is the best known for beer.

Cumming then resolved in his own mind in what manner he might get the same ingredients from Badenoch over the pathless hills which lie between the two countries; as there was no road, it seemed tedious, nay, almost impossible, to procure a ready and continued supply. Upon surveying the ground, he thought it might be practicable to make a road, and he resolved upon the arduous undertaking. He drew a line from Kilmavonaig through the woods of Craig Urrard, crossed the Bruar by Rièchlachrie, and so on almost in a straight line till it reached Gaig in Badenoch. He hired men, and made a road as far as Cum-na-feur, where the work was terminated in the following singular manner :—

There was a man at Ard Ghaith at Moulin named Mac Connoig, whose wife was a witch, and she resolved, with the assistance of another witch who lived in Croc Barrodh, a small village near her, to put a stop to Cumming's Road by their infernal magic; they metamorphosed themselves into the form of eagles; for those who are in familiar alliance with Clootie obtain from him the power of transformation. The Atholl tradition says, "it is not known whether Cumming ever injured them, or whether they bore him any malice or ill will." But the Badenoch history declares that these eagles were the transformed mothers of the girls whom he had commanded to reap stark naked on the following day.

Whatever they were, however, they took their flight till they came to Cumming's workmen, and by some charm they dispersed the men, and put the horses and oxen to flight, till they were driven over a great precipice, which was then called Cum-na-feur, or the Cart's Precipice.

Cumming, affrighted at the catastrophe, took to flight and galloped off, pursued by the two winged witches; he did not, however, acquit himself as well as Tam O'Shanter, for his body was torn from his horse by the eagles, the flesh stripped off, and nothing remained in the stirrup but one of his legs. The horse stopped for a space on the banks of the Tarff; and the spot where he paused is still called Lechois (one foot).

Thus terminated, according to tradition, the extravagant speculation with which Cumming was to supply Badenoch with Kilmavonaig beer; the length he proceeded with his work may be easily discerned at the present day; no person doubts that there was a road.

The belief in witches, fairies, and other supernatural powers has very much decreased of late years in Scotland; but it is a great mistake to consider it as wholly extirpated. Those who come in contact with passing strangers will naturally be reluctant to confess any superstition, for fear of being derided; but such as live in the country, and have free intercourse with the cottagers, well know with what deep reverence they relate such stories as these. They have descended from their ancestors, and they regard them as part of their creed. In a family in Atholl where there is now an old man residing, many of the long winter nights are spent in telling stories about ghosts, fairies, witches, warlocks, etc., which are solemnly listened to, and most religiously believed; and should any one of the company attempt to discredit these stories, or to try to account for them on natural principles, the hoary sage would treat such incredulity with ridicule, and regard the person as a most infatuated sceptic.

There is great talk of a witch that still haunts Ben-y-gloe. She is represented as of a very mischievous and malevolent disposition, driving cattle into morasses, where they perish, and riding the forest horses by night, till covered with mire and sweat, they drop down from fatigue and exhaustion. She has the power of taking the shape of an eagle, raven, hind, or any other animal that may suit her purpose. She destroys bridges, and allures people to the margin of the flood, by exhibiting a semblance of floating treasures, which they lose their lives in grasping at.

This very formidable person, in conjunction with the hunt given to **James the Fifth**, gave rise to the following legendary tale, which was given me in manuscript at Blair. It was written by the late M. G. Lewis during his visit there, and I am not aware that it has ever appeared in print:—

THE WITCH OF BEN-Y-GLOE.

I call thee! I charm thee! wing hither thy way!
By the laws below that the fiends obey!
By the groans which shall rise at the Judgment-day,
I call thee! I charm thee! wing hither thy way!

She heard him on her mount of stone,
Where on snakes alive she was feeding alone;
And straight her limbs she anointed all
With basilisk's blood and viper's gall.

But seeing, before away she sped
That her snakes half-eaten, were not yet dead,
She crush'd their heads with fiendish spite,
But had not the mercy to kill them quite.

Oh! then she mounted the back of the blast,
And sail'd o'er woods and waters fast;
She stopped on a rock awhile to rest,
And she throttled the young in an eagle's nest.

And now again her flight she takes
O'er rocks and muirs—o'er hills and lakes:
She saw below her the harvest swell,
And she groan'd to see that it promised so well.

She stops for a moment to curse the grain,
Then away on the wind she hurries amain;
Now she flies high—now she flies low—
And she lights on the summit of huge Ben-y-gloe.

Thither had call'd her a woful wight
With many a spell and mystic rite;
But when he saw the witch appear,
That woful wight he quiver'd with fear.

" Woful wight, now tell me true,
What hast thou summon'd me hither to do?
Woful wight, thy answer make;
I must be gone ere morning break."

" My son was a robber so stout and so bold—
Lo, where he lies pale, bloody, and cold;—
Revenge! revenge I ask of thee;
Oh! grant that Lord Atholl as cold may be.

" Atholl's earl, whose cup I bear,
Slew this morn my son so fair;
Though a robber he was, he was dear to me,—
So revenge! revenge I ask of thee."

" Now, woful wight, my counsel take,
And Atholl's blood thy wrath shall slake;
To work him harm three spells I know,
But more than three I may not show.

" These herbs of maddening power must feed
Ere dawn of day his favourite steed;
Then soon as Lord Atholl shall touch the reins,
Shall the steed dash out his master's brains.

" And if any one hears and dares betray
My secret ere St. Andrew's Day,
I'll drink his blood and crack each bone,
And turn the strings of his heart to stone.

" This cup did fiends at midnight make
By the heat of the burning brimstone lake;
In this Lord Atholl's liquor pour,
And if once he drinks, he'll never drink more.

" And if any one hears and dares betray
My secret ere St. Andrew's Day,
I'll drink his blood and crack each bone,
And turn the strings of his heart to stone.

" And should your foe these spells evade,
Then be the third and last essay'd;
Nor doubt I'll glut your vengeful spite
With blood, ere ends to-morrow night.

THE WITCH OF BEN-Y-GLOE

"For I'll hide you in Lord Atholl's room,
And wrap your form in magic gloom,
Till near his bed you can softly creep,
When your dirk may stab him while buried in sleep.

"And if any one hears and dares betray
My secret ere St. Andrew's Day,
I'll drink his blood and crack each bone,
And turn the strings of his heart to stone."

Sighing with sorrow, and burning with rage,
All this had heard Lord Atholl's page;
Who, curious, had follow'd the woful wight
When he sought the mountain's snow-crowned height.

Home he sped with heavy cheer,
"Oh how shall I save my master dear?
Oh how shall I manage the truth to tell,
Yet avoid myself the beldame's spell?"

Thus mourned the page till broke the morn,
But he sprang from bed when he heard the horn,
The jolly horn which, loud and clear,
Summoned King Jamie to chase the deer.

For now two days with Lord Atholl had been
King Jamie the Fifth, and his mother the Queen;
With lords and with ladies, a goodly show,
And all were lodged on Ben-y-gloe.

And there to welcome guests so great,
Lord Atholl had built a palace of state,
And all without 'twas covered with green,
And all within with silken sheen.

And there were all fashions of exquisite fare,
And tanks full of delicate fish were there;
And the King and his nobles had all as good
As had they been still at proud Holyrood.

Each day that King Jamie had passed on his grounds
Had cost Lord Atholl a thousand pounds,
Yet ordered Lord Atholl (his splendour was such),
That the third should cost full thrice as much.

The Earl he rose with the morning light,
And soon he met with the woful wight,
Who proffered a draught of cordial power,
To cheer his heart ere he left his bower.

Sweet Willie the page was at hand—the bowl
He knew, and terror seized his soul;
For he saw the Earl accept the gift,
And soon to his lips the cordial lift.

But ere they touched the goblet's side,
Sweet Willie the page, "Hold! hold!" he cried,
"And before you drink, to the Virgin pray
That her blessing may fall on your sport to-day."

The Earl then he sank on his bended knee,
" Mother of God, now hear," prayed he,
But scarce the words his lips could pass,
When in fragments flew the mystic glass.

Startled Lord Atholl in fear and surprise,
On the woful wight he fixed his eyes,
But his doubts to clear he may not stay,
For the monarch was mounted, and called him away.

"Sweet Willie, run, sweet Willie, speed,
And bid them bring my favourite steed."
His mouth all foam, his eyes all flame,
Snorting and prancing the black steed came.

But ere on his back Lord Atholl could bound,
He heard sweet Willie's bowstring sound;
Whizzing flew the trusty dart,
Nor stopped ere it pierced the black steed's heart.

Lord Atholl, his face was black with rage,
He struck to the earth sweet Willie the page;
" Now pardon, dear master," did Willie exclaim,
"I shot at an eagle, and erred in my aim."

Again Lord Atholl smote him sore,
And bade him see his face no more,
Till the Queen-mother prayed him his wrath to assuage,
And forced him to pardon sweet Willie the page.

Gay was the chase—all hearts were light,
Save Willie's, who dreaded the coming of night;
Gay was the feast, and gay each guest,
Save Willie, whose soul sad thoughts oppress'd.

When he heard his master laugh with glee,
Ah! little his danger he knows, thought he;
When he saw him wine in his goblet pour,
He wept lest his lord should never drink more.

But hark! what horn so loud doth blow
That it shakes the green palace of high Ben-y-gloe;
At the gate now stops a herald his steed,
And towards the King's table he passes with speed.

"To horse, King Jamie! to horse and away!
For the English are coming in martial array;
Your lands they waste, your people they slay,
Then to horse, King Jamie, to horse and away!"

Up started King Jamie, and summoned his court—
"Thou hast shown me, Earl, right princely sport;
But what thou hast heard the herald tell,
Commands me this moment to bid thee farewell.

"But thou, Lord Atholl, till morn must wait,
Then marshal your vassals and follow me straight.
Mount! mount! my nobles, for I'll away,
Though dark be the night, nor wait for day."

King Jamie is gone through mist and gloom,
And the Earl now seeks that fatal room,
Where the Witch, with blood to glut his spite,
Already had hid the woful wight.

But when on the lock was the Earl's hand laid,
"Alas! that the King," sweet Willie thus said,
"Exposed to the dangers of darkness should go,
But if *I* were Lord Atholl it should not be so;

"For rather of these towers I'd make
A bonfire for my Sovereign's sake,
Which, spreading wide its friendly light,
Should guide him safe through the dangers of night."

Lord Atholl, his head was hot with wine,
He heard and adopted sweet Willie's design,
He bade his vassals the palace forsake,
And each in his hand a firebrand take.

And he burnt the palace so stately and fair,
With hangings so rich and pictures so rare,
And with vessels of silver and vessels of gold,
And swift through the chambers the bright flames rolled.

But hark! who shrieks in pain and fright?
The fire has seized on the woful wight,
Who close in his master's room did lie,
And whom none had warned from the flames to fly.

And lo! while his life the miscreant ends,
On a column of smoke what fiend ascends?
'T is the Witch, who in curses vents her ire,
As scorch'd she flies from the raging fire.

All view'd the Witch in strange surprise,
But what she was could none devise,
Till St. Andrew's Day had come and flown,
Then made sweet Willie the secret known.

And he told, how thrice he had managed to save
His Lord, when he stood on the brink of the Grave;
And he told how his Lord had paid him with blows
For snatching his life from deadly foes.

Lord Atholl, he gave sweet Willie his hand,
And he gave him gold, and he gave him land,
And he gave him a wife, who was fit to be queen,
'T was his lovely daughter Gallantine.

Now if lords and if ladies are curious to know
What became of the Witch when she left Ben-y-gloe,
'T is right to inform them, for fear of mistakes,
That home she went, and finish'd her snakes.

CHAPTER VII.

Deer-drive to Glen Tilt.—Anticipated sport.—The deer-stalker's rhymes.—The start from Bruar Lodge.—Combat of stags.—Cautious exploring.—Stalking the great Braemar hart.—The shot and bay.—Preparation for driving the deer.—Dalnacardoc chamois.—A French sportsman.—The ambuscade, skirmish, and slaughter.—Shot at the black deer.—The party assembled.—The last hart brought to bay.—The bay broken.—The death-shot.—A carpet knight.—Condoling with a victim.—The Count's adventure.—Chase and capture of a poacher.—A quiet shot.—Granting a favour.—Termination of the day's sport.

"Ye shall be set at such a tryst,
That hart and hind shall come to your fyst."
Squyer of lowe Degre.

THE lord of the forest had now determined upon having a grand deer-drive to Glen Tilt, and Lightfoot was invited to make one of the party; thus, in a short time, this fortunate sportsman had an opportunity of seeing every variety and description of this interesting chase. That the show of deer might be as ample as possible, Tortoise had

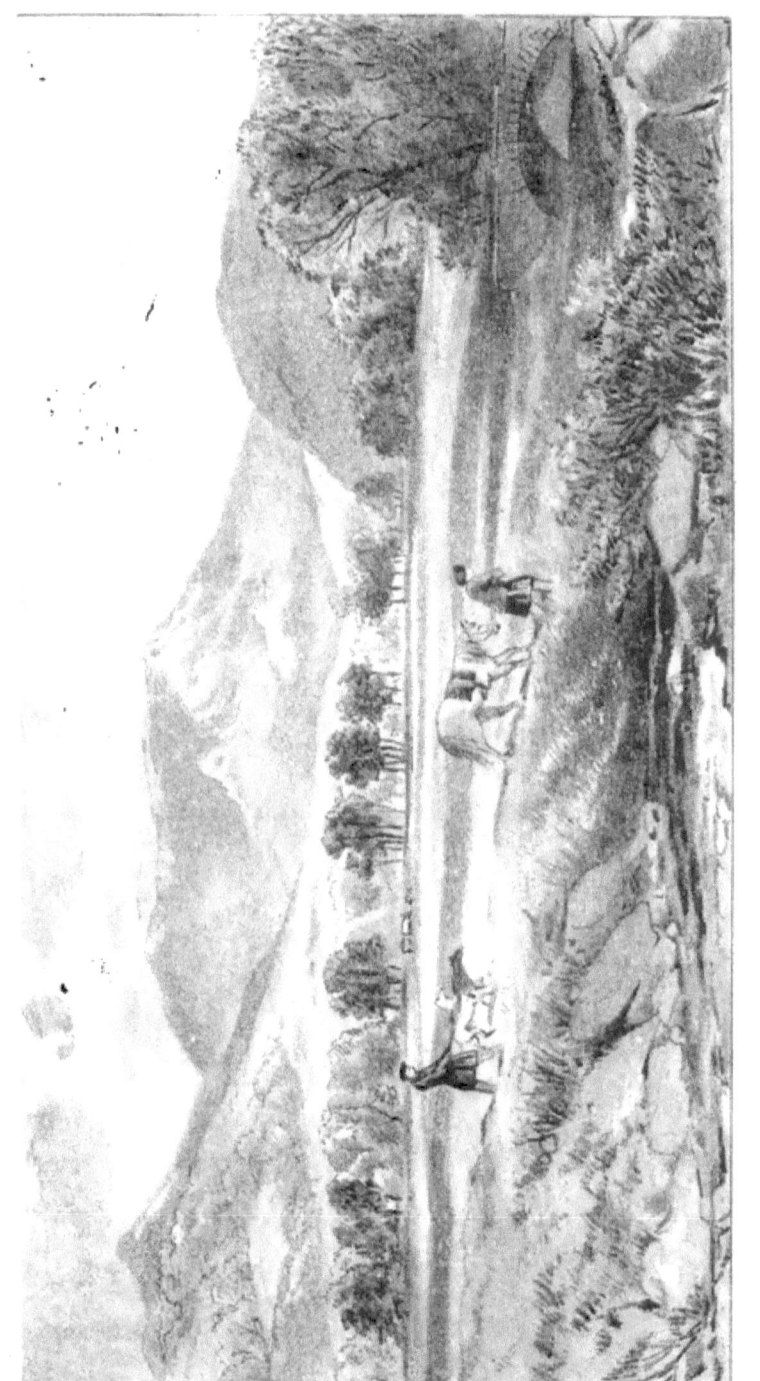

instructions to commence his cast at the remote parts of the forest, kill what he could, and get forward as many deer as he was able: he therefore dispatched all his men to Bruar Lodge over night, that they might be fresh and ready for the morrow's sport; a time was fixed for his meeting the foresters from Glen Tilt on Sroin-a-cro, when he and his men were to take the command of the right wing of the drive.

This animating sport was always enjoyed by anticipation; and you might easily read in the happy countenances of the guests at Blair, that something highly pleasurable and exciting was about to take place. When John Crerer and the foresters were summoned to the corridor over night curiosity rose to the highest pitch. Something positively awful was going on—was Glen Croinie to be driven, and would any one be suffered to go with the drivers? This great mystery was seldom solved over night; nor could it be so with certainty, as a change of wind must necessarily cause a change of operations. But on the destined morning each sportsman had clear and distinct instructions, and his proper station allotted to him; some of the old ones, however, who were knowing as to the currents of the air, and acquainted with the passes, were apt to finesse a little, and ingratiate themselves into the most favourable positions. These gentlemen might be seen, at the dawn of day, walking about the castle, and noting the precise direction of the clouds.

Modern hunting-parties in Glen Tilt, although not on so extensive a scale as those in days of yore, when nobles went forth with all their retinue, and the whole scene had as much the appearance of a military display as of a hunting excursion, were yet of a liberal, exciting, and lordly character. Parties of hill-men were sent forth, at a stated time, to form a semicircular line on the mountains, and press the deer down the crags into Glen Tilt, which they usually crossed, and then went forward, reeking and steaming, up the heights of Ben-y-gloe.

There were several stations in the glen, in which the various sportsmen were concealed, and from these no one was permitted to stir till the deer had fairly passed them.

These drives took place only when the wind was favourable, and, at such a time it was pretty easy to calculate at what hour the deer would come in sight.

It was not unusual for the drivers to collect a herd of five or six hundred head; and, occasionally, when they came down into the glen, broke into parcels, and turned back upon the drivers, the scene was splendid and animated, and the firing became very general; after the shots, dogs were turned loose, for the chance of bringing some of the fat sluggards to bay, and an excellent one it was.

Sportsmen, whose discretion and forbearance could be relied upon, were occasionally sent with the drivers, one at each wing, but it was their duty to consult the general sport, and not to get forward and fire, unless deer broke fairly out, lest they should turn the whole herd.

A scene so full of novel interest caused many a fluttering heart on the previous day, and many a feverish dream at night. Visions of deer, perhaps, came and vanished amidst broken slumbers; then the restless sleeper was lost and bewildered amongst mountains and torrents; then came a sudden start, as if falling from a precipice; lastly, and, oh, worst of all! an attempt to pull the trigger at a monstrous hart, without being able to effect the explosion of the rifle.

At length the shades of night pass away, and the morning breaks forth fair and beautiful.

THE DEER-STALKER'S RHYMES.

By the Hon. T. H. LIDDELL.

AWAKE and be stirring,—the daylight's appearing,
The wind's in the south, and the mountains are clearing;
A thousand wild deer in the forest are feeding,
And many a hart before night shall lie bleeding.

Make ready both rifles—the old and the new—
And sharpen the edge of the rusted skene-dhu!
Let your telescopes gleam in the rising sun;
We'll have need of them all ere the day's work be done.

The laddie was off before light to Glen Tilt,
And Fascally's laird has just tied on his kilt;
And Peter and Charlie are waiting below
The cloud-mantled summits of huge Ben-y-gloe.

Then spur on your ponies, and haste to the slaughter,
Where the Tilt and the Tarff mix their eddying water;
The ravens have spied us, and croak as they wheel
O'er the antler'd heads of their destined meal.

Now brace up your sinews, give play to your lungs,
Keep open your eyes, and keep silent your tongues;
And follow with cautious and stealthy tread
The forester's footsteps wherever they lead.

Here pause we a moment, while yonder slope
He surveys with the balanced telescope:
By heavens! he sees them—just under the hill
The pride of the forest lie browsing and still.

"Yon moss must be past ere we gain our shot,—
'T is full five hundred yards to the fatal spot."
So near has he reckon'd—that, as we crawl by,
Lo! the points of their horns on the line of the sky.

We have travers'd the flat, and we lurk behind
A rock, to recover our nerve and our wind:—
Hist! the calves are belling; and, snuffing the air,
Two jealous old hinds to the front repair.

See the herd is alarm'd, and o'er the height
The leading hinds have advanc'd into sight:
"Hold! hold your hand till the antlers appear,
For the heaviest harts are still in the rear."

Crack, crack! go the rifles,—for either shot
A noble hart, bleeding, sinks on the spot;
The third ball has miss'd,—but the hindmost stag
Was struck by the fourth as he topped the crag.

"Uncouple the lurchers!"—right onward they fly,
With out-stretching limb, and with fire-flashing eye:
On the track of his blood they are winging their way;
They gain on his traces,—he stands at bay!

Magnificent creature! to reach thee I strain
Through forest and glen, over mountain and plain;
Yet, now thou art fallen, thy fate I deplore,
And lament that the reign of thy greatness is o'er.

Where now is that courage, late bounding so high,
That acuteness of scent, and that brilliance of eye;
That fleetness of foot, which, out-speeding the wind,
Has so often left death and destruction behind.

Thine heart's blood is streaming, thy vigour gone by,
Thy fleet foot is palsied, and glazed is thine eye :—
The last hard convulsion of death has come o'er thee,—
Magnificent creature! who would not deplore thee?

Coir-na-Minghie has rung to the rifle's first crack,
And the heights of Cairn-chlamain shall echo it back;
Glen Croinie's wild caverns the yelling shall hear
Of the blood-hound that traces the **fugitive** deer.

By the gods, 't is a gallant beginning :—Hurra!
Diana has smiled on the hunters to-day!
In the sports of the morning come, goddess, and **share**,
And Bacchus shall welcome thee homeward to **Blair.**

The first who started for the sport were Tortoise and his men, of whom Jamieson was the chief—a fine, straight, sinewy, well-favoured man he was, with as good wind, as cool judgment, and as quick an eye for deer as any man on the hills. They had slept, as has been noted, at Bruar Lodge, about nine miles north of Blair, that they might begin at the outskirts of the preserved part of the forest. As soon as the morning mist was dispersed, they were breathing the fresh air on the summit of Ben Dairg, sitting upon the red stones, and prying with their glasses into every part of the vast forest that lay expanded before them,—more especially and minutely examining those places that were under the wind, the warm corries, and the best pastures. They had hitherto seen nothing but hinds; but, as such gear only spoil sport, they took care to give them their wind, and send them out north, that they might at once get rid of them.

It was now far on towards the rutting season; and, as the party advanced, and looked over the Elrich, they saw a parcel of hinds with a master hart, who had made this very Turkish collection for his sole individual gratification; these were to be kept, as they were obtained, by the strong antler. Like the Athenians in their prosperity, these martial fellows acknowledged no law but that of force.

Whilst the hart was walking proudly with the hinds, a hoarse roar comes over the ridge of the hill; it is the menace of war—nearer and louder it falls upon the ear; and, lo! the angry rival appears on the sky-line. He halts upon a projecting crag, swelling, jutting out his neck, and drawing himself up to his full proportions. Having now screwed up his courage to the sticking-place, he turns aside, and winds down the moss, bellowing and tossing abroad the heather with his antlers, his wrath seeming to increase as he moves onwards; his dauntless adversary sends back a loud defiance, and rushes forth to meet him in fair combat. The hinds wheel their ranks, and stand, with curious gaze and erect ears, to witness the joust,—and now the combatants meet brow to brow, butting and goring each other with great fury, till at length their antlers are fairly locked together. After some violent struggles they extricate themselves; and, being well matched, and quite exhausted, both sink upon their knees, and rest a space in that posture, still antler to antler. Somewhat revived by this brief cessation, they set to again, till the intruder, being at length forced backwards to the edge of a precipice, and feeling himself worsted, turns quickly aside, and fairly takes to flight, but runs in circles round the hinds, as reluctant to leave them. The victor follows close at his heels, goring him in the haunches; ever as he is touched he starts aside, till at length, beaten and jaded, he fairly gives up the contest, and gallops away, still hotly pursued.

Whilst this chase after the fugitive was continuing, in comes another hart from the opposite quarter; but no sooner had the victor heard his bellowing, than he returned to secure his hinds, and quickly drove this gay gentleman away, who took to his heels incontinently, being a beast of no mark or likelihood.*

After this amusing spectacle was over, these deer, being of no service to the drive, were suffered to go into Glen Dirie.

The party, now having ascended to the summit of Coir-na-miseach, crept forward cautiously behind a ridge of ground, and got a view of that immense basin called the Culreach. Instantly, as they looked below, there was a whisper of caution; they crawled back on their hands and

---

* This law of *detur fortiori*" is an admirable provision of nature for keeping the stock from degenerating.

knees, sunk the hill again, and posted themselves on safe ground. They had seen the deer, which were scattered up and down the hill sides, some grazing, others basking in the morning sunbeam, fat and lazy, whilst the jealous hinds were so disposed as to prevent any sudden inroad upon their position. Some of them kept to the wind, and others were continually looking towards those points from which they could not profit by it.

Jamieson now went back to take a minute inspection of the whole herd. He soon returned with an expression of eager excitement,—" There are several good harts," he said, " in the herd on the eastern face of the hill; but," added he, " there is a small parcel below us, and, as sure as deid, the great Braemar hart is among them—there is him and a small hart and five hinds a' thegither, and I'm thinking that he is so high up on the face o' the hill, that he may be pit over, and ye may hae a chance at him at last."

" Capital news, Thomas, and a glorious thing it will be if it should turn out so, for he is a hart of a thousand; but are you sure it is the muckle deer after all? The Braemar hart, who has foiled us twice, has a very sleek body, with high horns, not widely spread, and only eight points. You should know him well—are you sure it is him?"

" I could pick him out from a' the harts in the forest, and gie evidence against him, for he is a wary beast, and we have had sair work wi' him, he has led us mony a mile!"

Dispositions were now made for getting the herd forward into Glen Croinie; this was easily done, though it took up some time, for it was necessary to place a man towards the east, and another to the north, the sportsmen remaining on the western hill. These men soon arrived at their stations, and came forward at the concerted moment, working well together. So distant were they, that they could scarcely be discerned through the telescope. The herd soon took the alarm, and began to put themselves in motion. They drew closer together, the hinds gazed around them, and the harts, rising up from their lair, tossed up their antlers, and stood erect in their full proportion. As the hill-men advanced slowly and cautiously, the deer closed, and went forward leisurely; they then made a halt on the face of the hill, and formed into a beautiful group; but, as the drivers persevered, they drew out into a long string, and went at an easy pace up the steep towards Glen Croinie; arrived at the summit, they mended their pace, and each deer galloped over the scalp of the hill as if all the rifles of Atholl were at his heels, so that, in a few moments more, the whole herd were fairly in the glen.

There never had been the slightest doubt of the success of this operation: all Tortoise's anxiety leaned towards the small parcel which contained the great Braemar deer. When the general alarm took place these stood and gazed like the rest, and advanced some way as if to join them; till at length, when they made off, the proud leader stopped for a space, tossed up his antlers, and, disdaining to follow the servile herd, turned up the western face of the hill where Tortoise was lying: as he went forward the rifleman advanced also, preserving the wind, and just keeping sight of the points of his horns from over the brow of the hill.

The hill-men, seeing the favourable course he was likely to take, did their utmost to make him persevere in it. Every thing looked propitious; but still it was uncertain whether he would come out from the hollow at a favourable point of the hill, or go over the easy swell, where it would be impossible, from the nature of the ground, to come within distance of him; indeed, he seemed inclined to do the latter. What an anxious moment was this for the rifleman! who can tell what hopes, and what dire apprehensions shot rapidly across his mind, when he saw the pride of the forest almost within his reach? forward he came, bounding and pitching up the hill, casting his broad shadow on the green-sward, and followed closely by his companions. As yet, his course is dubious;—now he bears more to the west, and races along, as it seems, rather in sport than in fear;—by heavens! he nears the rifleman:—on for your life and make your push! With bent body, but with rapid steps, Tortoise ducked down, slipped suddenly back behind the eminence, and then went forward at the top of his speed. The horns, which he never lost sight of, are seen approaching the hill-top—down again crouched the rifleman for a moment, till the course of the deer was decided then another **swift** movement below the hill brought him within distance, **just as** the magnificent fellow had passed the summit, and **was** descending into the opposite glen.

Tortoise's breast had been in **a** tumult, but it was lulled in a moment—

> " Che sue virtuti accolse,
> **Tutte in quel** punto, ed in guardia **al cor le mise."**

He stopped suddenly, like a bolt **that had hit the mark ;** —stood firm—clapped his rifle rapidly **to his shoulder, and** fired just as the hart was disappearing **from his view.**

" Habet,—he has it—he has **it, Jamieson ; I heard the smack of** the ball **true enough."**

" **Hurra, he lags behind !** Now, then, let go Tarff: quick —quick, Sandy ; **lose not** a moment ; quick, for your life, man ; we cannot **wait till** he falls out : come here, Jamieson ; I, and my men must join the general drive, or the deer will **break out** ; so take you one of the rifles, and finish that fine **fellow as he** goes to bay in **Glen Mark :** you will have no **time to return,** so do **not attempt** to come back up Sroin-a-cro or Cairn-Marnoch ; **you will be more useful** in the glen **by** keeping the deer in on **that side.** You can come in at **Auk-mark-moor. Away with you."**

And away went the stout hill-man, bounding over moss and hillock, till in a few minutes he sunk down from the view.

" Hark ! **I hear** the baying of the hound : now **it dies** away :—**Do you hear it** now, **Sandy ?"**

" No, I did not ; I heard naething but **the corbie."**

" Look with your glass, **then, whilst I load."**

" Hey ! what a sight ; I never kent **the like o' it afore."**

" Why, what do you see, man ?"

" Why, sure the deer is chasing **Tarff** all ow'r the moss, and Tarff is rinnin awa' joost ahead o' him ;—I never kent the like. Now the hart stops—now **Tarff** is at him again : ah, take care, Tarff!—Now the deer has beaten him aff, and is rinnin after him **again."**

" I see it all myself, **Sandy, with** the glass ; and I see, too, that one dog, be he what he may, can never manage that deer ; so let go Derig, **for he has** heard the bay, and

will soon be up with him." And so, indeed, he was: glen and mountain now resounded with the raging of the deep-mouthed hounds, till at length the vexed quarry broke down the river Mark, and then, turning aside and skirting the Brae, stood before a huge mass of rock that was anchored on the mountain side: thus posted, he boldly faced his antagonists. Thrice did the ferocious Derig spring aloft in the air, and fly ravenously at his throat, and thrice was he driven back with unmitigated fury. Maddening with rage, the fine animal rushed forward, raking and stabbing with his antlers, and gave chase, in his turn, to his enemies. It was a novel sight to see the noble beast act on the offensive. The war, when it ceased on the side of the stag, was again renewed by the hounds, who, although wounded and bleeding, ever returned stoutly to the charge. In vain was the rifle at hand, for the dogs were ever springing at the throat, in the way of the ball. And now, see, the bay is again broken, and away they go, right up the steeps of Ben-y-venie.

"Here we can tarry no longer, for the Duke's men are approaching; but it matters not, for Jamieson will inevitably bring that noble fellow down, though he will give him some trouble, and perhaps occasion the death of my good hounds.

"Well, Peter Fraser, here you are at last: when are we to start the deer?"

"At one o'clock exactly; and a' the men are round towards the east, under Charlie Crerer's command: then there's George Ritchie the fiddler at Cairn-y-chlamain; and Macpherson will gang doun Glen Croinie. The Duke trusts to you and yer men to pit ower the deer from the wast."

"Well, Peter, this is all as it should be, and the left wing cannot be under better command than that of Charlie Crerer; for, besides being a very clever fellow, he is as active as the beasts themselves, and always zealous to do his duty: a great regard I have for him, for he was my first instructor, and many a pleasant day we have had together in bye-gone times. As for the main body of the deer breaking on our side, we will so deal with them, that they shall not have that crime to answer for; if a few harts alone take such a fancy, my nature is not so cruel as to baulk them of their intent, since in that case I shall get a shot or two without prejudice to the general sport; if therefore this should happen, we will conduct ourselves with liberality, and suffer them to take their own pleasure without let or hindrance; and now, whilst we are waiting here, you may as well tell me what sport there has been at Dalnaspiedel.'

"I didna hear aething anent the moor-fowl at Dalnaspiedel; but I heard that the English gentlemen killed five deer at Dalnacardoch."

"Five deer! Deer at Dalnacardoch? How could they possibly come into deer in such ground as that?—What clever fellows they must be!"

"And clever chiels they were sure eneuch, for they got intill them without fashing themselves much aboot the matter; but the gentlemen, some gait or anither, had not studied nature, so that when they brought hame the beasties, the guidman at the inn couldna agree wi' them in opinion, though he is a very civil man too; for Sandy said that the five deer were five goats, whilst the gentlemen said that the five goats were five deer; but, sure eneuch, they had all beards, were wee beasties, and smelt like goats all over."

"Well, Peter, and how did all this end?"

"Why at last, then, they (that's the deer-stalkers) began to think that Sandy was richt, and that the deer were goats; so they behaved very handsome, and gave the farmer a hantle o' siller for their day's sport, being sorry for the mistake they had made: and it's mair the pity they didna prove to be deer; but it's no that easy to turn the like of an old goat into a fine hart."

"Well, Peter, I do not think that the sport was so bad after all; for I believe that the chamois, in chase of which the Swiss risk their lives, and are out for days together on mountains of eternal ice and snow, is little better than a great goat after all."

"I didna hear of sic a beast mysel; but I ken, by yer honour's account, he is no worth the speering at."

The moment had now arrived for starting the deer; and the signal was given, that every one might go on in good order, and act according to the movements of the quarry. Macpherson, who was to go down Glen Croinie, was instructed to keep in the rear till the deer were on the eastern face of the hill above the glen;—prudently did he hold back, for they were endeavouring to break out on the west: Tortoise and his men, however, turned them without difficulty; and, after some hard running and considerable manœuvring, they took precisely the desired direction.

But the drive, upon the whole, did not proceed with the usual alacrity; there was a sportsman (so called by courtesy) upon whose pace the hill-men on the east waited, and it was unfortunately a slow one; he had several shots, which were so injudiciously taken, that the success of the general sport seemed to be in jeopardy: the deer, I believe, were in none at all.

"Why, now, Peter, what in Heaven's name can that apparition be? Take your glass, and see what like it is."

"I see the man plain eneuch, for it is nae wraith; but I canna joost say what like he is, for I never kent the like o' him afore; he's nae Scotchman, and he hasna the tread of an English, for he aye gangs forrat on his toes wi' a wee bit jerk. Haw, haw, haw, I never saw sic a dress on the hills: do tak a gliff o' him through the prospect, yer honour."

"Ah, I see him, Peter, and I guess he is a Frenchman; but, with all his capering, he is as slow as a soldier marking time. Merciful he is, for not a beast has he touched as far as I can see. Surely he must be firing with blank cartridge; but the deer are going right in spite of him, so I hope he enjoys himself; but, at any rate, if he spoils sport in one way, I am sure he shows enough in another. I wonder what he thinks he is doing?"

And now the stately herd began to crown the summits, and were soon descried from the glen, hanging on the sky-line in long array. Those in the van gaze steadily on all sides,—onward move the others in succession, their horns and bodies looming large against the sky. Heavens! what a noble sight; how beautiful, how picturesque! See how they wind down the crags, with slow and measured steps; now hidden, and now reappearing from behind impending masses of rock: now the prudent leader halts his forces and closes up his files; those in advance are scrutinising the glen, whilst the rear-guard, wary and circumspect, are watching the motions of the persevering drivers. As the men come forward in a vast semicircle, the herd begin to mend their pace,—calves, hinds, and harts, come belling along, and wind down the oblique paths of the steep, putting in motion innumerable loose stones, that fall clattering over the crags.

The glen wore the appearance of utter solitude; but the sportsmen were lying in ambush in various parts of it, under the impending banks of the Tilt, behind fragments of rock, or in some cleft or position which screened them from the gaze of the deer.

And now how many bosoms were throbbing at this splendid spectacle, and what fitful changes from hope to despair agitated individual sportsmen, as the herd approached, or deviated from their particular position.

Beset upon their flanks and upon their rear, and seeing no obstruction in the wild forest before them, after long and deep misgivings, they take their desperate resolution: down they sweep in gallant array,—dash furiously across the meadow, and plunge right into the flashing waters of the Tilt. Hark how their hoofs clatter on its stony channel! Onward they rush,—the moss-stained waters flying around them, and are fast gaining the opposite bank.

Their course being thus decidedly taken, the lurking rifle-men spring up at once, like Clan Alpine's warriors, and rush forward on all sides.

Those who were fortunate enough to be near the spot of crossing had fair chances; for though some of the herd were cut off and turned back to the west, yet so long a string passed across the glen, that they had time to fire, and load, and fire again.

Many rifle-men came in, breathless, from more distant stations; some in time, and others all too late. Several shots were fired in distance, and out of distance, with various success; and the skirmish for a short space was pretty brisk on all sides. The herd, having fairly crossed the rocky channel of the Tilt, scampered away at a prodigious rate, and went forward, reeking and steaming, right up the face of the great mountain.

"Quick, quick, uncouple the lurchers."

The dogs spring from the leash, strive and press forwards; but are half blown before they come up with them. The herd now collect into a dense mass, each deer wedging himself into it as he finds he is the particular object of attack. Not a single hart fell out; and the hounds at length returned, with slinking countenances and drooping sterns,—lolling out their tongues, they lie panting on the greensward.

The sport however had been excellent; the Duke of Atholl (always the most skilful and successful of the party) killed three first-rate harts; our friend Lightfoot two, decidedly: two more were killed, as your rustic grammarian has it, somewhat promiscuously; and the old sportsmen also did considerable execution, selecting their harts with great tact. Moreover there were slain three hinds, that nobody would own to, and an exceedingly promising young fawn, repudiated also by all.* The French Count, whom we noticed on the mountains, distinguished himself in his own particular manner: but his high achievements well merit a separate history; and that they shall have.

And now let us go back to Tortoise, and see if he was idle all this time. No, not so; for a few harts and hinds broke over to the west, and, as the general sport was already secured, he used his pleasure with them. He had only two rifles, the third having been given to Jamieson to kill the deer at bay; he came well in to them, and, at the first shot, slew a noble hart; but there was another in the parcel still superior, which had been running on the opposite side: as the men got forward, the little herd came sweeping round over the open ground, towards Clashtyne, describing the segment of a circle.

---

* It was considered a disgrace, as has been elsewhere intimated, to kill hinds and fawns; a stranger, not aware of this, wrote to thank the late Duke of Gordon for a day's deer-shooting in Gawick; intimating how happy his Grace would be to hear of his success, for that he had wounded a hind, and killed an exceedingly promising young fawn.

"Lord! Lord! that black deer:* hey, what a deer! There, there, that black deer! Ou, he is ower far."

The words were scarcely out of Peter Fraser's mouth, ere the shot resounded in the hills. The hart was running swiftly, at about one hundred and fifty yards distance, or "by 'r Lady," somewhat more, but quite clear, and the ball seemed to smack against the centre of his body.

"Sandy, Sandy, the dogs, the dogs,—quick, quick, man! Lord, will ye never come forrat? Let go Shuloch. Here, Shuloch, Shuloch."

Away went the gallant old hound, upon his traces.

"And now he is safe enough; and we will leave him to Jamieson, who will meet with him at bay, as he comes down Glen Mark, where he will assuredly go. So leave Sandy to gralloch and bleed the other deer; and let us keep on down the hill, in case the great herd should be turned, and endeavour to come back over Auk-mark moor. They went over the ridge, however, in beautiful style, their backs all reddening in the sunshine; and they must, and will, cross the glen if every one keeps concealed till the right moment. Hark, I hear a shot! Another, and another, —glorious! Come along, Peter, skim down the mountain like a swallow: surely some of the herd will turn back upon us. There, there—Charlie, Crerer is running like an ostrich. Ah, Charlie, Charlie, it winna do; they are fairly past you, and you will pass us too, but not without a shot."

One rifle, in fact, was discharged by Tortoise as they swept by, and one more hart lay plunging in the heather.

"Now, then, let go Percy and Douglas after the others; and we shall send down a deer or two to the Tilt, which will make a noble day's sport.

"Bravo, Percy, bravo! See, he has taken out one hart, and Douglas another; they are sinking the hill, right down to the Tilt. Sit down whilst I load, and listen to the bay. I hear it sure enough now; it is Percy's bay. How he makes the valley ring; I should know his deep tongue from a thousand. He must be just above the marble quarry. Hark! that is his death-shot, and from the Duke; for no

---

* Black from rolling in the mire.

one else would fire at a deer at bay whilst his Grace was in the glen. We shall soon know this, for a few minutes will bring us within sight."

And now, as they bounded down the brae, the whole line of carriages, gillies, and sportsmen, broke full upon their view. That glen, heretofore so still and silent, awoke at once into life and animation. A large party had collected round Marble Lodge, and made a most picturesque appearance. Here a successful sportsman came, triumphantly galloping upon a mountain pony; and, far in his rear, riding at a dejected pace, loitered some unhappy wight, whose balls had been somewhat too busy with the heather. The wild gillies, soiled and heated with toil, were running to and fro in their blue bonnets and plaided kilts, some leading the good deer-hounds in the leash, with panting sides and flagging sterns; others, with fresh dogs, trotting lightly along, and looking up the mountain to the right and left, with keen gaze and half elevated ears. Nobles and kerns were mixed, and talking together with that good fellowship and equality, which a common interest in an animated pursuit so generally and so happily occasions. Or, if there was any ascendancy (always setting aside the Lord of the Forest), it was vested in John Crerer; so true it is, that "it is place that lessens and sets off." He was the Belarius, to whom the noble sportsmen looked up with deference and respect.

Three stout ponies, with redundant manes and shaggy coats, came slowly winding down the glen, each with a magnificent deer corded on his back. Tortoise had gone rapidly forward, with a fresh dog and a hill-man, in quest of Douglas and the deer: faintly he has heard the bay; now it peals louder and louder, as he rounds the wooded promontory.

"Now, speed thee, speed thee, Sandy; quick to the Duke, and tell him we have a noble hart at bay; this torrent and these cliffs he himself cannot gain, but say I will break the bay, and get him down to the Tilt, where he shall surely die the death. Lose not a moment, for time presses. Nay, never go round by the bridge, man,—the river, though swollen, is still fordable here, and will not wet you above your waist; plunge through at once. Well done, stout Sandy, you bear yourself like a true man."

"Time, indeed, was waning fast, for it was long since the birchin-leaves had trembled and glittered in the sunbeams, and the golden splendour, which so lately slept upon the mountain-top, had already died away, consigning it to its own stern and rugged nature. The air was coming up the glen, dank and chill; hill, brae, wood, and precipice were beginning to mingle in one universal melancholy mass.

The hart had got into the river Mark, just above the spot where it comes brawling into the Tilt; it was one of those deep chasms where the sunbeam never enters; in most places the rocks dropped steep, smooth, and shelving down to the flood. There were huge blocks of granite in the channel, and it seemed wonderful how the vexed animal could have got into the dark chasm in which he stood. But there he was—the torrent at his feet, and the long bony arm of a blasted birch stretched over him. Douglas stood baying at the point of a rock above, venting his vain wrath, and making stoops as if he would plunge down from that "bad eminence," but, sensible of his danger, he as often drew back; various were the attempts he made to come in at some other place, but still he was obliged to return to his first position. Tortoise now came up with Croinie; she was a most sagacious animal, and it was her custom to do the thing as coolly as possible, always running before the deer till she came to a convenient part of the river, when she turned in and headed him. This method she took in the present instance, and was soon swimming before his antlers.

But the stout animal would probably have remained immoveable in his position till the hound perished with cold and fatigue, had he seen no other enemies—indeed, he seemed to hold both dogs in thorough contempt; but when Tortoise stood before him, for a moment he raised up his stately crest, and waved his broad antlers to the right and left, gazing restlessly around him, then plunged at once down the torrent, trampled upon the hound, and bounded out far below, gaining the open birch-copse that skirted the banks of the Tilt. The hounds kept on their course, following

him through all his windings. Arrived at length at the
steep banks of the river, with one brave bound he gains the
centre of the stream; and there he stands majestic and firm,
and in ready act to do battle. The hounds dash after him
as best they may; fain would they attack him, but in vain
they stretch their powerless limbs: lightly does the hart
regard them, as they are swept to and fro by the rapids, and
can scarely hold their own. But when he finds the taint in
the air, and discovers sterner foes, he looks forward to the
free mountains before him, and again, breasting the flood,
strives every nerve to gain the heights of Ben-y-gloe. But,
alas! it may not be, Douglas and Croinie gain fast upon his
traces; and, after a rapid wheel, he plunges again into the
Tilt, and stops not, rests not, but down, down he goes,
through pool and over cataract, swimming, wading, and
rushing onward through the divided waters. The dogs,
close upon him, are borne down at times by the weight of
the flood, but rise up again to the surface, undaunted and
eager in the pursuit.

At length, and almost at the departure of daylight, the
Duke comes forward with his good rifle; one shot from that
unerring hand, an echo dying away through the mountains,
and see the fleet limbs fall powerless, and the dun carcase
goes floating down the stream, welling out the life-blood.
The current bears it onward rapidly, jostling against the
rocks, and wheeling in the eddies. In dash the kilted
foresters in gallant style, stemming the flood, and stretching
forth their arms in vain; their daring was perilous, girt
round and oppressed as they were with the waters; but still
the deer bore past them, always just beyond their reach.

But who is this coming forward with the ropes and
grappling-hooks? Who but the excellent and trusty
M'Millan,* mounted on his sheltie, "and charming the
glen with fair feats of horsemanship." Gently now, my

---

* John M'Millan entered the Duke of Atholl's service in 1791, as assistant-fisherman to Duncan Kennedy. When Duncan died he became the principal fisherman. He was a powerful man, and a most valuable and attached servant; but never could acquire skill in the ways of the deer. He rode so awkwardly that he seldom mounted a pony without getting a fall. He was unfortunately drowned in the Tay, near his own house, in January 7, 1836, at the age of 71 years.

feathered Mercury, I pray and beseeech you not to swerve so undecidedly to the north and to the south, but resolve me at once towards which point of the compass you mean to make your summerset; for your pony, mark me, is a recusant, and, sooth to say, I never saw any animal less solicitous of another, than that beastie is of his rider. There now,—hope you're not hurt. Pick him up, Charlie, and take on the grappling-hooks to yon pool; you will get the hart out easily there, for he will sweep round in the cheek of the stream.

Out he was taken triumphantly, and there he lay on the green sward, bausy and sleek, "the admired of all admirers." Some praised his beautiful form, and held up his widespreading antlers; whilst others (not oblivious of currant-jelly) began to handle him after the fashion of Parson Trulliber. Certain it is he enjoyed great posthumous fame. But here comes Jamieson, hurried and heated with toil.

"Well, Thomas,* have you finished that great devil?"†

"Yes, I got him on Ben-y-venie, where he went last to bay. But both dogs are wounded: Tarff not much; but Derig, you see, is stabbed badly in four places; and I doubt he may not recover."

"Ah, poor fellow, what terrible wounds he has in his chest and loins; that in his side is not so bad, for I see the horn has only passed between his skin and his ribs. Well, my brave Derig, you shall go home in the cart, and be carefully looked after. And the great black deer, Jamieson, that Shuloch took into Glen Mark; did you get him?"

"Quite easy; he was shot through the body, and made but a poor bay."

"Capital; we have made clean work of it, then, at last."

"Joy, joy to you, Lightfoot; they say you have killed two first-rate harts: what a happy mortal you must be! But do, pray, tell me who that smart foreigner is who so nearly spoiled all our sport."

---

* Thomas Jamieson lived formerly at Abbotsford, and came into the author's service many years ago with Sir Walter Scott's permission. He now acts as principal gamekeeper, and is in every way a most valuable servant.

† The author has kept the horns of this deer, which are splintered at the points, by coming in contact with the rocks when the dogs escaped from the thrust.

"Most readily will I give you his history, partly collected from the hill-men, and partly from my own observation; for when his grand affair took place I heard and saw all."

"He is a French noble, who has had the merit of bringing himself into notice as a famous shot; not, as I conceive, from any feats of skill that he has actually performed, but simply from his excellent *soi disant* qualities. He is, as you see, beautifully equipped; that, indeed, no one can deny; dressed, too, in the most elaborate style. See how knowingly his rifle is slung in the German fashion. I assure you that, what with his gay good humour, and foreign singularity, he has attracted a considerable degree of observation: 'His discourse is sweet and voluble;' but aged ears by no means 'played truant with his tales;' for John Crerer and the older sportsmen discovered properties in him quite adequate, they said, to destroy the sport of a whole season. What was to be done? If he remained in the glen, it was imperative on him to be totally silent: singing French airs was out of the question. The deer, said the Duke, were not to be had as in the time of Orpheus; on the contrary, it was more becoming to be mute, and to lie concealed like Marius in the marshes of Minturnæ, and somewhat better. But it seemed quite evident that nothing short of the combined powers of laudanum and a straitwaistcoat could effect any restraint upon our gentleman. These were not at hand, and, if they had been so, it might perhaps have been thought somewhat inhospitable to have used them; so that idea was dropped at once. In this dilemma it was deemed advisable to send him up with the drivers, to plague you: in short, it was resolved that he should evacuate the glen. He started joyfully, for he was a famous walker—out of all sight the best in France; indeed no one of any nation was equal to him. But the hill-men asserted that this was not his particular walking-day; so that, I am told, he soon became most deplorably exhausted, and, according to all accounts, delayed the drive at least an hour or so. Fortune bounteously gave him many fair shots; but, alas, what she distributed with one hand, she took away with the other; for he missed them clean every one."

13

"*Mais c'est étonnant celà.* I who never make the miss!"

"Perhaps your honour forgot to put in the ball."

"*Ah! voilà ce que c'est, vous l'avez trouvé, mon ami. Le moyen de tuer sans balle!* Now, then, I put in the powder of cannon, and there goes de ball upon the top of it *mort de ma vie!* I now kill all the stag in Scotland, except a leetle, and you shall surproise much."

He was a bad prophet, for he still went on, missing as before, amongst winking hill-men and grinning gillies. At length, however, the sun of his glory (which had been so long eclipsed) shone forth in amazing splendour. "Fortune," says Fluellen, "is painted upon a wheel, to signify to you (which is the moral of it) that she is turning and inconstant, and mutabilities and variations:" and the turn was now in the Count's favour, for she directed his unwilling rifle right towards the middle of a herd of deer, which stood "thick as the autumnal leaves that strew the brooks of Vallombrosa." Everything was propitious: circumstance, situation, and effect; for he was descending the mountain in full view of our whole assemblage of sportsmen. A fine stag, in the midst of the herd, fell to the crack of his rifle. "Hah, hah!" forward ran the Count, and sat upon the prostrate deer triumphing. "*Hé bien, mon ami, vous êtes mort donc! Moi je fais toujours des coups sûrs. Ah! pauvre enfant!*" He then patted the sides of the animal in pure wantonness, and looked east, west, north, and south for applause, the happiest of the happy; finally he extracted a Mosaic snuff-box from his pocket, and, with an air that nature has denied to all save the French nation, he held a pinch to the deer's nose: "*Prends, mon ami, prends donc.*" This operation had scarcely been performed, when the hart, who had only been stunned, or perhaps shot through the loins, sprang up suddenly, overturned the Count, ran fairly away, and was never seen again.

"*Arrête toi, traitre, arrête, mon enfant. Ah, c'est un enfant perdu! Allez donc à tous les diables.*"

Thus ended the Count's chasse. Everybody was very sorry, and nobody laughed, of course; as for me, by my troth, I will never follow Frenchman's fashion in deer-stalking.

"Capital! our Parisian friend beats the Italian gentleman, who exhibited in the forest of Glengarry, all to fits; though this latter noble was also of a joyous turn, and a complete contemner of Harpocrates. He was posted, as I have heard, at a deer-drive in one of the best passes, with strict injunctions as to concealment; unfortunately, he made a slight mistake between the letter and spirit of the law. It is true he hid his own person very skilfully, but placed his bonnet aloft on a birch branch (the weather being hot), in rather a commanding situation; at length, feeling somewhat solitary, he began to awaken the echoes by singing Italian airs—

"Eurydice, the woods, Eurydice, the floods,
Eurydice the rocks and hollow mountains rung."

It is needless to tell you that his Excellency had what is termed "a blank day."

Although a numerous herd of deer had been brought down, something had evidently gone wrong towards the east. The Count had for some time the merit of this failure; indeed, if he did not totally spoil the drive it was not his fault. There was another hero, however, who shared the honours with him. It was evident that a large parcel of deer, which ought to have come down, had got the wind of some one on the moor; the point and manner of their starting was marked by the hill-men, and two of them, suspecting foul play, went forward to examine the ground. One of these men held a lurcher in the leash. After exploring the moor for some time, they came to a deep ravine. Still they saw no one. But, in following its course a little way, the lurcher held back, stretched forth his neck, and gave a low growl. At this unequivocal sign, one of the men clambered down, and discovered a hind that was newly killed; and, as he was advancing under a projecting mass of rock, out bolted a kilted man with a gun in his hand: hot pursuit immediately commenced. The poacher went right up the chasm, down which fell a considerable quantity of water; the man, who had ascended, followed after him, whilst the other who held the dog remained at top, ready to cap him when he came out. The pursuit was close and hot; the poacher cutting out good work; the pace, however (owing to the nature of the ground), was a bad one in itself. But who can make rapid way up a stony cataract? They scrambled and splashed, and fell forward on their hands, and cut their shins, climbing over masses of rock that were lying in the channel, and jumping from them into the water-course, till, at length, the Duke's man got hold of the delinquent's foot as he was just springing from a rock above him.

"Haud him fast, Donald, man—dinna lowse him—dinna let him gang awa at ony gait."

"Nae fear ava. The de'il a man ever got frae me when it aince cam' to close grups."

He was completely mistaken, however, for the fugitive slipped out his foot, and left his shoe only prisoner, which the wrathful hill-man sent at his head, accompanied by some thundering Gaelic anathema. And now the poacher dropped his gun; perhaps he meant it as a lure, like the fabled golden apple; or, perhaps, it was done to favour his speed. However this may be, he certainly made much better play without it.

When Donald, who was on the banks of the ravine, saw how things were going on, he took the advantage of the even ground, headed him, and then came down into the chasm in front of him, so that our worthy friend was placed between two fires.

Thus have I seen two cunning terriers hem in a poaching fox: they rage and press closely upon him, whilst the woods and mountains ring with their shrill clamour. Meanwhile the insulted beast, fixed in a position from which he sees no escape, bears his brush against a rock, shows his white teeth, and commences gallant defensive warfare. Not so our hero; he thought little of deeds of arms;—of fine and imprisonment much. The spectre turnkey was before him, and ugly visions of high grated walls and solitary dungeons made him desperate. With the vault of Grimaldi he seized hold of an impending branch of birch, swung himself aloft by strength of arm, and fairly escaped from the abyss, leaving his enemies gazing up from below. He got a capital start, for the hill-men could not extricate themselves with the same alacrity.

The bay being now broken, they had a beautiful race over the moor; but the light-limbed foresters gained ground; the fugitive's pace became worse and worse; he laboured and floundered, and was at length seized, all breathless and exhausted.

"Why, how dare the like of ye to come intill his Grace's forest, and steal his deer; ye shall pay the lawin, man?"

"Hout-tout! I'm nae thief ava; it's joost for my ain diversion; but ye hae bin owre muckle wi' the Southrons; and the like o' thae chiels aye ca' liftin', stealing."

"I think I ha' heard that afore," says Donald. "What! my friend the Gown-cromb of Badenoch? will no the Lias-mor, or great Garden o' Eden content the man? must he come stalkin', and feeling the deer in the braes o' Atholl?"

The notorious blacksmith was soon taken down to Glen Tilt, and brought into presence of the Duke of Atholl: after a sharp remonstrance, his Grace asked him whether he would go to Perth gaol for three months, or stand a shot from his rifle at a hundred paces.

The man said he would stand the shot.

"Very well;—John Crerer, step out a hundred yards."

The ground was measured.

"Now post the man with his front right towards me, and give me my best rifle, John."

The gun was given, and raised slowly, whilst the hill-men stood by in a group in breathless suspense; the direction of their eyes changing alternately from his Grace to the man. A long and steady aim was taken—it was an awful moment, but the blacksmith neither flinched nor stirred; at length the cap of the rifle only exploded.

"Pshaw! Give me another rifle, John, and take care that it be better loaded."

The second rifle missed fire also, as well it might, it having been of course arranged that there should be no charge in it.

"Well, you are a lucky fellow, for I see your time is not yet come. Give the man his fill of whiskey, John; he does not lack courage: but mark me, Master Gown-cromb, if ever you come after my deer again, my rifle will not miss fire;

and if it does, the gaol at Perth is large enough to hold you, and all the poachers in Badenoch, though ye are a numerous progeny."

"I winna say that I will gang entirely wi'out my sport, for I canna aye be wanting venison; but yer Grace shall never find me in yer forest again. There's mony a stoot hart in Glenfiddock, and mony a yell hind in the pine woods of Braemar, let alone Gaig and Glen Feshie; and I will leave the braes of Atholl for yer Grace to tak yer pleasure in, and never fash them more since ye request the favour."

Thus ended a deer hunt, fit for the recreation of King Jamie; and although stags were not slain by hundreds, as Lesley has chronicled, or by scores, as the water poet has recounted—both of which accounts I hold to be gross exaggerations—yet the sport probably was quite as ample in proportion to the numbers engaged in it, and the small space of time that was occupied in bringing down the deer.

The glen, too, as in times of yore, was graced by the presence of many a fair and noble dame who had been waiting the termination of the drive in the mountain lodges; indeed, it is recorded that ladies of high station have not only felt a great inclination towards this noble sport, but have actually engaged in it.

"Her Majesty" (Queen Elizabeth), says a courtier, writing to Sir Robert Sidney, "is well and excellently disposed to hunting; for every second day she is on horseback, and continues the sport long." At this time she was in her seventy-seventh year, and was then at her palace at Oatlands.*

The party now proceeded to the hospitable halls of Blair; where we will leave them, amidst cultivated society and high-born beauty.

" To fight their battles o'er again,
And thrice to slay the slain."

---

*There are various other notices of the delight this Queen took in the diversion of killing the stag.

CHAPTER VIII.

OF POACHERS AND FREEBOOTERS.

Forest Contracts.—Wandering Poachers.—English Vagabonds.—Adventure at Felaar.—Highland Vampire.—Peter Breck's Backsliding.—Trap Baited with Whiskey.—The Gaig Pet Stolen.—Poacher's Adventure.—Desolate Situation.—A Highland Witch.—Chisholm's Cave.—Freebooter's Life.—John More.—Sutherland Monster.—A Priest in Jeopardy.—Highland Robin Hood.—Ourna-kelig.—The Widow's Hospitality.—Rival Poachers in Atholl.—Adventure in Glen Tilt.—Rob Doun.—Curious Trial for Murder.—A Polyglot Ghost.—Ghost no Lawyer.

> " Donald Caird can wire a maukin';
> Kens the wiles o' dun deer staukin';
> Listers kippers, makes a shift
> To shoot a moor-fowl in a drift.
> Water-bailiffs, rangers, keepers,
> He can wauk whiles they are sleepers;
> Nor for bountith or reward,
> Dare you mell wi' Donald Caird."
>
> SIR W. SCOTT.

THE passion for hunting wild animals is probably one of the most powerful affections of the mind where it has once taken root. It is the recreation of nobles and of kings, the solace of the gentry, and the allurement of the paradise of wild nations. After death, the Indian of the West believes that he shall ascend the Rocky Mountains, " and there among the crags, and snows, and tumbling torrents; and, after many moons of painful toil, he will reach the summit, from whence he will have a view over the land of souls: there he will see the happy hunting-grounds, with the spirits of the brave and the good, living in tents in the green meadows, by bright running streams; or hunting the herds of buffalo, and elks, and deer, which have been slain on earth." *

If we look back to remote periods in our own country, we shall find that the most severe laws—mutilation, and even the penalty of death itself, have not had sufficient terrors to arrest the course of unlawful depredators. Deer-killing by poachers was formerly carried on to such an

---

* Adventures of Captain Bonneville, by W. Irving, Esq.

extent, that the proprietors of forests found it necessary to combine in order to protect their mutual interests. In the "Collectanea de rebus Albanicis" of the Iona Club, there is a contract, dated November 3, 1628, between several of the principal Highland lairds "for the preservation of deer and roe on their respective estates, and the punishment of trespassers; mutually binding themselves to respect each other's forests, and cause them to be respected by their retainers, under special penalties, according to the rank of the person transgressing: a hundred merks for a gentleman, with forfeiture of the hagbute or bow; £40 for a tenant; and, in case of a common man, "his body to be punishit according as pleises the superior of the forest: ane witness sufficient." They appear to have had a sort of jury trial for poachers.

There are several old Acts of the Scottish Parliament "anent steilors of hart, hynd, roe, and doe, to be punishit as thift, and anent shuitteries at Thame; quhilk is appointed to be punishit with *death*, and escheit of their gudes moveable." These laws have been reckoned barbarous, but they are not more severe than those which, in former times, were in force against sheep-stealers, taking likewise into consideration, that sheep are of infinitely less value than deer. If it be true that deer wander from one forest to another, so that no laird can claim a certain property in them, it is also obvious that the common poacher can have no right in any case, and must steal from some one or another. The claim can rest with the landed proprietors only. It is a fair give and take business according to the direction of the wind; your third man, however, steps in, and I think enjoys the sport much more than those who are privileged to follow it. In the "History of Badenoch," it is mentioned that Cluny Macpherson deprived a man of his arm, and of one of his eyes, who killed deer afterwards in this mutilated condition. I do not mean to defend the lawless proceedings of poachers, but I cannot help confessing that there is something so adventurous, and so full of picturesque character in these rough fellows—so much skill exhibited by them, and such endurance of climate and fatigue, as may in some degree be admitted as extenuating qualities; and I would not, as Shakspeare's town-clerk says, "condemn them to everlasting *redemption* for this;" I would simply transport 'em to the wilds of America, where they could enjoy the sport without injury to any one, and we might carry on a trade of furs and skins with such free trappers.

Deer-poaching is carried on in two distinct methods. In the one case, by a man who belongs more or less to the spot, and who hovers about the moors, watching the keepers, and seizing his opportunity; and in the other, by gangs of marauders who go from forest to forest, as the wind serves, and act in concert. These latter men take possession of some deserted bothy, or even of the shooting-lodges, if they are left unoccupied. There was a bothy on Tarff side so frequented by them, that it was thought necessary to pull it down altogether. These poachers commence their operations chiefly at the termination of the regular season, so that the harts are entirely rank and useless. The yeld hinds, however, come in at that period, and are very fine venison; and all the other hinds make the best possible soup, and are very good hill-man's eating, though they are totally devoid of fat.

Such poachers as go about in gangs are rough, picturesque-looking fellows, able to face any weather; and they act, as I have said, in concert. Being a stronger force, they can remove the deer which they kill, without much inconvenience, and can readily dispose of it in the country. During the time of their depredations, they subsist upon what is not saleable; and with this, and their whiskey and tobacco, they must pass a very pleasant wandering life. It is extremely difficult for keepers to apprehend these foragers, as all of them have glasses, and cannot easily be surprised in the open country. The best way is to attack the bothy by night, and a fine animating scene it would be. I do not think that the men would endeavour to save themselves by the commission of murder. They have still a high reverence for their chieftains, which would restrain them from committing bloody excesses upon their lawful agents. In the Highlands, one never hears of such ruffians as infest the preserves in England; men who screw up their courage at the beer-houses, asserting with imprecations that they will shoot any keeper rather than be taken. A vicious set they are, bringing up their families in idleness and profligacy; proceeding from crime to crime, till at last their career ends either on the gallows, or in transportation. I have fined and imprisoned scores of these vagabonds, some of them two or three times over, and I never yet heard of one that was reclaimed. They are absolute thieves, for there can be no sport in taking a hare out of a wire, or shooting a pheasant on his perch by night.

Your Gael, on the contrary, has a fine rough sort of sense of honour about him—peculiar enough to be sure—thus, "the man who refused thirty thousand pounds for betraying his prince, was hanged at last for stealing a cow." It was not long since a poacher was taken in the forest of Braemar: having some good points in his character, the nobleman who rents the ground very generously told him, that if he would promise never to poach again in that district, his gun, which had been taken from him, should be restored, and he himself should be set at liberty. He very coolly replied that he wished to have an hour to consider of the matter; at the expiration of that time he stepped forward and said, "Ye may tak' my gun, and me too, for I will no gie the promise."

Occasionally some superstitious dread will do more to prevent deer-stealing than the most rigid legal enactments. An instance having such a tendency occurred some years ago in the forest of Atholl.

There is a shooting lodge built at Felaar, which lying between the Atholl and Braemar country, has often afforded a warm night's rest to travellers overtaken by darkness in that bleak and rugged country; when left unoccupied, it has frequently been taken possession of by poachers. Two such characters arrived there some few years ago after a pretty successful foray, and finding the door resist their efforts, they broke open the window, and thus having gained admittance, they kindled a fire, and began to consider themselves quite at home. Their first object was to prepare their supper, but having no water in the house, one of them undertook to fetch some; for this purpose he was obliged to get out of the window. Having put forth his legs first, he was resting his arms on the window-sill, with his face fronted to the interior of the cottage. Whilst in this position, he began to scream with all his might, roaring out that some fiend had hold of his leg, and was tearing it and sucking his blood. At length, by a violent struggle, he extricated himself, and gained the ground, still in great fright and pain. In searching round he could see neither man nor beast, nor any living thing. But he very gravely asserted that he saw some white objects and some faint blue lights at a distance, which continually shifted their situations, and at length vanished entirely.

Having procured water, he did not venture to return through the window; but the door was broken open by the united efforts of himself and his companion. They spent the night in a state of superstitious alarm, nor could they on the following morn discover the track of man or beast about the place; their own footsteps alone were visible. The injury remained for a considerable period; the man, indeed, bore the marks of it all his life, as many people now living at Blair can testify. This occurrence, remaining unaccounted for, had such an effect, that no poacher took up his quarters at Felaar Cottage in after times.

Men of this description usually set forth at night when the keepers have retired, that they may be on the desired ground betimes in the morning; thus they gain some hours upon them. If the wind serves, their first manœuvre is to get the deer out of the forest, which is very easily done; and when they have them there, they keep them as long as they can; but unless they go clear away to another forest they generally return by a circuit with a side wind at night. The only method to defeat these lawless proceedings, is to throw up peat bothies near the outskirts of the forest at proper intervals, and place keepers in them. Such men must be constant in their residence, or the poachers will exchange places with them.

I will now relate a story which shows that the keepers themselves had not in former times a very nice perception of equity:—

In the month of July, 1783, the Duke of Atholl summoned his three principal foresters, John Crerer, Moon, and Peter Robertson, and promised a handsome reward to him who should kill the fattest hart within the allotted period of two days, which was meant as a present to the king (George III.) Crerer and Moon set forward on the following morning before daybreak, each attended by a hill-man, and provided with a horse. Not so Peter Robertson, better known by the name of Peter Breck (from his being pitted with the small-pox). He had revolved a scheme in his mind which required privacy and craft worthy of the best times of Johnny Armstrong. A sort of raid it was, or lifting from his neighbours' grounds—that is to say from the lands of Gaig.* These lands were at the time possessed by Stewart, of Garth (the late General Stewart's father), and another gentleman; they kept their sheep in Gaig all the summer and during the harvest, and on a low farm in the winter and spring. Alexander MacDougall and Archibald MacDermid were shepherds in Gaig for many years; and they had taken a fawn,† which they tamed, and brought up with two milch cows that were pastured in Gaig all the summer; and at the time I am now treating of, this pet hart was five years old. He was taken to the low farm during winter and spring, and generally lodged every night in the barn; they fed him upon oats, hay, barley, or peas in the straw, of which latter provender he was extravagantly fond. By these means he became enormously fat, and of a towering size, so that he probably exceeded in weight any hart in the forest of Atholl. Now Peter Breck was mindful of this bonny beast, and had often turned the tail of his eye upon him; but his virtue, or, it may be, the manner in which the animal was guarded, had hitherto borne him out against all temptations. That virtue, however, so impregnable when little was to be gained, began to succumb before the promised reward. Great allowances must be made for our friend Breck's backsliding, for lifting was not quite disgraceful in those days; besides the animal was fat, stupendous in size, and, in short, altogether undeniable. So Peter took his sheltie and attendant, slunk away cannily in the gloaming, proceeded up Glenbruar, and

---

* Spelt also Gawick.
† Calf is the proper term, but both are used.

arrived, at the grey dawn of day, at the shepherd's lodge at Gaig. He had previously left his attendant and his horse and gun a considerable distance above the lodge, at a place called Gargaig. He soon roused the shepherds from their slumbers, and, pretending to be very drunk, laid himself down upon one of the beds they had quitted. This was all very natural, for Peter had no great character for sobriety; loud and deep did he snore—never surely was sleep so sound.

And now, as he was lying dormant, as it seemed, what should the shepherds see but the black neck of a whiskey bottle peeping out from one of his pockets. Why should they not tak' it? What for no? the man was fou already, and couldna want mair. Out it came then, and was soon despatched. The said bottle was then filled with water, and returned to the place from whence they extracted it. Breck then turned restlessly on his other side, when, lo! the neck of another bottle delighted the eyes of the fortunate herdsmen; this was treated precisely in the same manner as the first had been, for Breck's snoring was awful, and they were safe enough from interruption. As soon as this second bottle had been filled with water and replaced in the pocket, Peter thought proper to awake. The shepherds now having drunk a bottle of whiskey each, had little inclination to go to the hill; so they made a fire, and began to cook some victuals; Breck joined them as they were eating, and told them he could help them to some good whiskey, which he had in his pocket: this they thought it prudent to decline, saying it was too early to drink; but little suspecting that he had been watching all their motions.

Both the herdsmen soon became heavy, and feeling inclined to sleep, the one threw himself on the bed, and the other slept on his seat by the fireside. Breck having thus far accomplished his object, stole out of the bothy, and seeing the cows and the stag browsing in the plain below, he drove them slowly to Gargaig, where he had left his rifle, horse, and attendant. The stag followed the cows, as he was accustomed to do; and now being fairly at too great a distance from the lodge for his shot to be heard, he levelled, and despatched the hart most deliberately. No time was lost in cording it on the horse, and off he went homewards as fast as he could; but the horse, although a good Highland garron, had such difficulty in carrying his heavy burthen, that they were obliged to rest at Glenbruar, and it was dusk before they reached the castle of Blair.

Breck's arrival made no small sensation; the Duke hastened out to see what he had brought home; and being surprised at the great size of the animal, which was brought to the portal of the castle, asked where he had the good fortune to kill it. "Not on your Grace's grounds," was the reply,—"Where then?" inquired the Duke. "On the Inverness-shire hills," replied Breck: "I have had this hart in my eye for years, and have seen him frequently, but never in the company of any other deer." On being weighed, he was found to be nineteen stone, Dutch weight, without the gralloch.

Breck got the reward, somewhat to the mortification of Moon and of Crerer, who were better men. The truth, however, soon broke out, and his competitors lost no time in reporting to the Duke that Breck had stolen the Gaig pet. His Grace sent for him, and demanded if it were true that he had stolen it. Breck denied the theft lustily;—he 'couldna say' but that it was the Gaig pet, but declared that he had got it from the shepherds for a Scotch pint of whiskey, which is about two quarts. The Duke expressing his surprise that they should part with it for such a trifle, Breck explained to his Grace, that the shepherds were aware that he (Breck) knew that they had got the stag, when a fawn in the Atholl forest; as well as that they frequently poached both deer and moor-fowl there; so that, under these considerations, they gave up the pet for the Scotch pint. Peter, however, had still to reckon with the shepherds; but he held their attack lightly, and told them, that they were repaid tenfold by their depredations on the Atholl forest, thanked them for the care they had taken of his fawn, and advised them never to steal an honest man's whiskey again, taking advantage of his being asleep.

In the year 1773, two poachers set forth from the Braemar country in quest of deer; the weather had been lowering for some time, and when they arrived at Tarff

Side, they were overtaken by a snow storm; it was not however severe, and when it cleared up, the wind being north, they soon got a parcel of deer out of the forest of Atholl: these made a long start, as they always do when the wind is in that quarter; thus the men had them quite away from the preserved part of the forest, and in a situation where they were not likely to be interfered with.

After considerable manœuvring, which occupied the greater part of the day, they wounded a hind, and traced her a long distance by her blood-drops on the snow. In the meantime, as the day drew near a close, the wind rose, and the snow-blast returned with greater violence; and having been intent on following the traces of the wounded deer, they had wandered about till they were completely lost. In this condition they heaped up a few stones and turfs, and having their plaids, and some oat-cake and whiskey with them, passed the night without any very serious inconvenience.

The dawn brought no alleviation to their anxiety; the winds howled, and the snow fell, so that no outline of mountain or landmark could be seen. It was now no longer a question of killing deer, but of saving their lives. The wind, which continued north, was their only guide, and by turning their back upon it, they avoided the brunt of the storm, and had hopes of reaching Glen Tilt or the Strath of the Tay. The snow had drifted in such masses, that they were unable to pursue any decided line, and it was so deep in all places where the wind had not acted upon it, that their advance was very slow and laborious.

The small stock of provisions which they took out with them was exhausted; the wind got more into the east—a change they were not aware of—so that in turning their backs upon it, they travelled towards the west instead of towards the south, as they fancied they were doing.

At length, when night was setting in, they saw a deep and unknown glen of joyless aspect before them; they descended into it, to avoid the bleak winds of the summits, and had proposed to put up a few stones and turfs for shelter during the dark hours. Whilst they were looking for a convenient spot, to their great relief they discovered a shieling, deserted, as they imagined, as buildings in such remote places usually are in the winter. What, then, was their surprise, when, upon approaching the door, it was at once opened, even without their knocking. A woman presented herself, of a wild and haggard aspect; told them she had been expecting them, and that their supper and beds were ready. Even so they found it—the pot was boiling, and bannocks and oat-cake were placed upon the table, and also two plates, for the expected guests. There was something so extraordinary about this old woman, that it operated as a sort of fascination, and the men's eyes were continually turned upon her. She had large features, long lank hair, and small grey eyes, deeply sunk, and conveying a striking expression of vice and cunning; she halted on one leg, and chaunted a wild song, in an unknown language, while she was pouring out the kail.

Tired and exhausted as the men were, the whole thing appeared to their superstitious imaginations so much like witchcraft, that, although half famished, they could scarcely bring themselves to eat. Fear came upon them, when she waved her long sinewy arms, and darkly hinted that she had power over the winds and the storm, muttering at intervals some unintelligible sentences; then at once holding up a rope, with three knots tied in it: "If," quoth she, "I lowse the first, there shall blaw a fair wind, such as the deer stalker may wish; if I lowse the second, a stronger blast shall sweep o'er the hills; and if I lowse the third, sic a storm will brack out, as neither man or beast can thole; and the blast shall yowl down the corries and the glens, and the pines shall fa' crashin' into the torrents, and this bare arm shall guide the course o' the storm, as I sit on my throne of Cairn-Gower, on the tap of Ben-y-Gloe. Weel did ye ken my po'er the day, when the wind was cauld and deidly, and all was dimmed in snaw,—and ye see that ye was expectit here, and ye hae brought nae venison; but if ye mean to thrive, ye maun place a fat hart, or a yeld hind in the braes of Atholl, by Fraser's cairn, at midnight, the first Monday in every month, while the season lasts,—the laird's ghaist will no meddle wi' it. If ye neglect this my bidding, foul will befall ye, and the fate of Walter of Rhuairm shall o'ertake ye; ye shall surely perish on the waste; the raven shall croak your dirge; and your banes shall be pickit by the eagle."

Awed, superstitious, and depressed as they were by fatigue, the poachers were not backward in giving the promise, though it is not very probable that they ever performed it. They passed the night in deep sleep, and it was late before they rose from their beds of heather, when they asserted that their hostess had vanished.

The snow storm having ceased, they found their way into the track which led to Blair, and got into the strath of the Tay. This is supposed to have been the last time that the witch of Ben-y-gloe held converse with mortal man; but those who were less given to superstition believed that the woman had been expecting her own friends, who were probably also poachers detained by the storm, and that she had made use of the above artifices in order to obtain venison.

Chisholm's Cave, in Carn-Vaduc, in the Ben Klibreck forest, in Sutherland, derives its name from a freebooter, who passed his life in caverns, poaching and living upon pillage. His early history cannot be traced satisfactorily; but it is probable that he became a recluse in consequence of having committed some atrocious crime; and that he selected the retired cave at the back of Klibreck, from his love of a forest life. He was not a native of Sutherland, nor had he, whilst there, been guilty of any heinous crime; but he scrupled not to make frequent nocturnal visits to the inhabited parts of Strathnaver, and, on such occasions, to carry off to his caverns, corn, and such other necessaries, as were not to be procured around his desolate abode.

The large cave, which bears his name, is an extensive winding cavity, or rather a succession of open spaces, or holes of unusual size, such as Brobdignag rabbits might be supposed to haunt. In this dismal labyrinth, Chisholm lived many years; it is said he kept two cows underground, and left venison in lieu of the hay and grain which he plundered in the cultivated strath.

This sort of bartering gave little offence; nay, some were gratified by it, for Chisholm was dreaded as a lawless man, whom it was dangerous to anger or molest: they considered that a person who could live in the gloomy holes under Carn-Vaduc, must be in the service of the powers of darkness, and that it was not safe or canny to interfere with him. Even the foresters used to shun him, though he was never known to offer personal violence. He lived so much apart from the rest of mankind, and was so seldom seen, that his dress and appearance became latterly a matter of doubt, and the manner and time of his death was never known. He either removed privately from the country, or expired in one of the remote chambers of the cavern, which no person was hardy enough to explore.

A similar system of free living was adopted by a man named John More, who lived in Durness about the same time, and rented a small farm near the Dirrie-more. He neither had, nor cared to have, permission to kill deer and game; but his whole time was devoted to poaching, and his wild mode of life rendered him an uncouth but tolerated plunderer of the forest.

Donald Lord Reay happening to pass near John More's residence one summer morning, determined to call and endeavour to reclaim him from his lawless propensities. He left his attendants at some distance, that he might ensure confidence on the part of his rude host. He found John at home, and told him that he called to get some breakfast. John was evidently proud of this visit, and pleased with the frank manner in which he was accosted, having been usually threatened by those in authority with imprisonment and the gallows.

"Come in, Donald," said John, in Gaelic, "and sit on my stool, and you will get to eat what cost me some trouble in collecting."

His lordship entered the hut, and was soon seated in a dismal corner; but John opened a wooden shutter that had filled up a hole in the wall, through which day-light entered, and revealed a tall black-looking box, which was the only article in the house that could be used as a table. John bustled about with great activity, and, to his lordship's surprise, pulled out from the box two or three beautifully white dinner napkins. One of them was placed on the top of the box as a tablecloth, and the other spread on his lordship's knees. The fire, which glimmered in the centre of the room, was then roused, and made to burn more freely. This proceeding denoted that John had some provisions to cook;—from a dark mysterious recess he drew forth a fine grilse, already split open and ready for being dressed. By means of two long wooden spigots, which skewered the fish, and the points of which were stuck into the earthen hearth, the grilse was placed before the burning peats, and turned occasionally. Soon after a suspicious-looking piece of meat was placed over the embers; and when all was cooked, John placed it upon the box before his chief, saying—" John More's fattest dish is ready :"—adding, that the salmon* was from one of his lordship's rivers, and the meat the breast of a deer. Lord Reay asked for a knife and some salt; but John replied—" that teeth and hands were of little use, if they could not master dead fish and flesh; that the deer seasoned their flesh with salt on the hill, whilst the herring could not do so in the sea; and that the salmon, like the Durness butter, was better without salt."

John produced, also, some smuggled brandy; and pressed his lordship to eat and drink heartily, making many remarks on the manliness of eating a good breakfast.

The chief thought this a good opportunity to endeavour to make a proper impression upon his lawless host; and, after having been handsomely regaled by plunder from his own forest, determined to act with such generosity towards More as would keep him within reasonable bounds in future.

"I am well pleased, John," said he "that although you invade the property of others, you do not conceal the truth, and that you have freely given me the best entertainment that your depredations on my property have enabled you to bestow. I will, therefore, allow you to go occasionally to Fionavon in search of a deer, if you will engage not to interfere with deer, or any sort of game, in any other part in my forest."

More could never tolerate any restraint, and his answer was begun almost before Lord Reay had finished his handsome offer.

---

*A grilse is supposed to be a young salmon.

"Donald," said he, "you may put Fionavon in your paunch,—for wherever the deer are, there will John More be found."

This conversation was in Gaelic, in which language the peculiar phraseology is more piquant than can be rendered in English.

Donald MacCurrochy Mac-Ean-More, who lived latterly at Hope, was another very noted poacher in Sutherland. Numerous anecdotes are told of this man; but they refer rather to the great enormities he was in the habit of committing, than to his lighter trespasses amongst the deer. His acts of violence and injustice were so unusual and savage, as to render him an object of universal abhorrence.

His family name was Macleod. He deliberately murdered his nephew, that he might possess himself of the adjoining lands of Eddrachilles; and he afterwards put to death several of his friends, whose revenge he anticipated. He was an expert archer;—so ruthless a villain, and so ready to slay any one that offended him,—and, indeed, every one whom he could attack, whether friend or foe, that, at a period when the law was quite inoperative in the remote corners of the Highlands, he became the terror of the entire country. The greater part of his time was spent in the Dirriemore forest, where he was very successful with his long bow.

His nephew, when attacked by him, took refuge in a straw-covered hut, in an island on an inland loch; but MacCurrochy tied burning pitch and tow to the head of an arrow, and firing it into the roof, set the place in flames. The young man endeavoured to escape by swimming, but an arrow from the ruffian's bow pierced his heart just as he was reaching the shore.

MacCurrochy's shieling was without a door or window, and he entered by a hole in the roof, from which he would occasionally take a shot at a passing traveller. It is reported of him, that when walking with his son, a mere boy, on the banks of the river Hope, they saw a neighbouring priest on the opposite side of the river; young MacCurrochy exclaimed—

"O, daddy, give me your bow that I may bring down the priest."

"He is at too great a distance from you," said the father, "and you would get us into trouble, if you attempted to kill him without succeeding."

The priest, unconscious of his danger, approached nearer the river, and seated himself on a projecting stone.

"Now, daddy," said the youngster, "give me the bow, as I am certain I can hit him."

But the old man, still doubtful of his son's success, and expecting to obtain a nearer aim, refused this second request also. When the priest moved off, the boy insisted upon being permitted to shoot at the stone upon which he had been sitting; and having hit it with an arrow the very first trial, MacCurrochy complained bitterly of his want of judgment in having resisted his son's desire, and d——d himself "for vexing the boy's spirit."

MacCurrochy was master of a gun, which, along with his bow, he is said to have thrown into a deep cavity amongst the loose blocks of stone on the side of Craig-na-garbat, which forms a shoulder of Ben-Hope, when he felt himself dying. Many attempts have been made by the neighbouring inhabitants to discover these relics, but without success.

This ruthless villain was buried in a hole in the wall of Durness church, by his own direction, to baulk the threat of an old woman, who told him when he was dying that she should soon have the pleasure of dancing over his grave. There is a rude monument over his resting-place, on which a grotesque figure of Donald is cut, in which he is represented as drawing his bow and killing a deer. There is also an inscription, bearing date 1623, the year of his death. It runs as follows:—

> "Donald Makmarchor
> Hier lyis lo vas. il to his
> Friend, Var to his Fo:
> True: to his Maister in Veird
> And Vo."

Which was probably meant to pass as rhyme, thus,—

> Donald M'Marchow here lies low,
> Was ill to his friend, war (worse) to his foe;
> True to his master in word and vow,
> (Or in weal and woe).

Several of the forest anecdotes in Sutherland refer to a person known by the name of Our-na-Kelig, who resided in the parish of Loth, and who appears to have been not only a most successful and constant hunter of deer, but also a most stout and valiant clansman. His history is involved in considerable mystery, but his memory does not appear to have been tarnished with anything like secret assassination, or other serious crime. His proper name is unknown; that of Our-na-Kelig, by which alone he is referred to in tradition, is, I am told, descriptive of the grey, or light colour of his dress, and of his being a great eater of cod fish, or often engaged in catching it.

In a bloody skirmish between some Strathnaver men and those of the eastern coast of Sutherland, at Drumderg, in Glen Loth, Our-na-Kelig engaged one of the Strathnaver men, whose two sons also were present. He always laid about him with a two-handed sword, swinging it around with great fury, and letting it fall on his adversary with irresistible violence; giving such a stroke as Ariosto describes, when he says, "*Cala un fendente:*" Anglice,—" Lets fall a cleaver." With this formidable weapon he soon despatched the Strathnaver man,—whether or not he divided him from head to foot into two equal parts, tradition does not say; but it relates that the sons of the slain man rushed instantly on the victor with desperate rage, but only to meet the death of their father.

The Strathnaver men were defeated; and the fame previously acquired by Our-na-Kelig as a formidable swordsman, was prodigiously increased by the slaughter of three powerful men in open combat.

Soon after this onslaught, Our-na-Kelig went into the Ben Ormin forest to kill himself some venison, as he was wont to do, without being very particular about the laws of property.

"———— The good old rule
Sufficed him; the simple plan,
That they should take who have the power,
And they should keep who can."

He bent the best and the stiffest bow in the country, killed a deer when he was hungry, and would devour a whole limb of it, hastily roasted between two peat fires, lighted for the purpose on the open heath. Well, he set forth in quest of venison, nor had he been out long before he wounded a hart, and sent his dog after him. The chase led him far away over the hills, and he was overtaken by a heavy snow-storm; benumbed with cold, and weary with floundering in the drift, his only hope for preserving his life consisted, perhaps, in being able to reach one of the shielings in Strathnaver. After long and painful toil—his life-blood chilled, and in a state of dreadful exhaustion—he arrived after nightfall at a small bothy during one of the most bitter blasts of the storm; far different now in plight than on that memorable day when he signalised himself in combat, he humbly sued for shelter. The shieling was inhabited only by a woman and her daughter, who, being intimidated, refused his request. He earnestly answered that he was so worn out by struggling against the storm, that he could go no farther, and that he must shortly perish if refused admittance. The poor woman's kind heart got the better of her fears, and she removed the fastenings of the door; then, as it was driven inwards by the violence of the wind, and as the snow beat upon her careworn face, she said in Gaelic, whilst the tear stood in her eye—

"That on such a night as this she could not refuse admission into her bothy even to Our-na-Kelig himself, should he be wandering on the moor, although he had slain her good-man and her two brave sons, and left her ill to do in the world, and desolate."

Our-na-Kelig was not personally known to this poor widow, and having obtained admittance and shelter, forebore to distress her feelings by revealing his name to one who had so much reason to dread and detest him. He ate of her meal, and restored his benumbed limbs before her peat-fire; and it may be that his heart smote him as he felt his vigour returning, and cast his eyes upon his wretched preserver. He parted from her next morning with expressions of gratitude; and upon his return home, sent her five bolls of meal from his rich corn farm in the parish of Loth, and continued the same gratuity to her annually during her life.

The following account will prove the extent to which poaching was occasionally carried on, even in the face of honest and vigilant keepers.

One of the most notorious poachers in Atholl forest in former days, was D—— S——.* He resided in the district, and is still living. He kept his masons upon venison whilst they were building his house in 1812, and subsisted his family on the same diet.

This D—— S——, accompanied by C—— R—— and A—— O——e, went forth on a poaching excursion in the forest of Atholl, when they knew that the regular foresters were upon distant duty. After having killed two or three deer, which C—— R——, as being the least skilful shot of the party, was left to gralloch, night came on, and they boldly made for the lodge of Ridorrach.† Early the next morning the wind was to the north, and they saw a long string of deer coming forward towards Ben Derig; judging from this that some one was coming up the glen, they shifted their quarters without loss of time, crossed the Bruar, and from an eminence on the west of the river, with the help of their glasses, they spied Donald Macbeath, the forester, who lived down the glen, at Richlachrie, and who was coming up the water side.

Having the advantage of the ground, and the wind being north, this did not impede their operations, and by one o'clock D—— S—— had shot two hinds. In the midst of this success they saw three men (who had probably been stalking them for some time) running towards them at full speed. They immediately took to flight, but having their guns and other incumbrances with them, whilst their pursuers were empty-handed, they lost ground rapidly. Thus they were reduced to the predicament either of surrendering or giving battle. Things being in this state, O——e motioned to them with his hand to keep back, and told them that if they did not mind his voice, he would send a stronger and a more unwelcome messenger to them. They

---

* Some of these worthies being still in existence, their names are partly suppressed according to promise.
† I am not quite clear as to the accuracy of this name, being unacquainted with it myself.

paid no heed to this threat, and O——e, actually putting his cowardly threat into execution, levelled his piece and fired. The ball struck the snow at the feet of one of the party.

S—— and R——, his comrades, were thunderstruck at this mad act of O——e, and peaceably awaited the coming up of the other party, who proved to be poachers like themselves. The man fired at was outrageous, and he had good reason to be so; but after various threats on one side, and submission on the other, matters at length took a pacific turn.

These poachers who had given chase, finding that others of the fraternity were before them, and were putting a complete obstacle to their success, hid their guns, and endeavoured, by passing for keepers, to drive them out of the forest. The finesse, as has been seen, wanted but little of ending in bloodshed.

When men went forth singly on these unlawful excursions, they were sometimes placed in considerable difficulties for want of efficient assistance. A poacher had very lately a desperate struggle in Glen Tilt, the particulars of which I mention as they came from his own mouth, for he was never discovered.

He set off in the evening, that he might be on a deer-cast by the grey of the morning: whilst it was dark he descried the horns of a deer in a hollow very near him; he had small shot only in his gun, and was in such a position that he could not change the charge without danger of disturbing the stag. He crept, however, so close to him, that when he sprung on his legs, he fell to the shot. Not a little surprised, the poacher threw down his gun, dashed forward and seized his victim by the hind leg; but it was no easy matter to hold him. In this struggle the man kept his grip firmly, whilst the deer dragged him at a tearing pace amongst the large stones and birch hags, till he was all over bruises, his legs severely lacerated, and his clothes torn to shreds; his bonnet and plaid had entirely disappeared.

He now contrived to get hold of his knife, but it dropped in the struggle; and as the deer still sustained its vigour, he had much ado to keep hold of the limb even with both his hands. The darkness became deeper as the animal tore and strained forward, through the skirts of a birch wood, and both repeatedly fell together.

Breaking forth again into the open moor, he found his weight was beginning to tell on the energy of the stag, so that he had power to swing him from side to side, till at length, just as they were re-entering the wood, this determined bull-dog of a fellow fairly laid him on his broadside, and with such force, that the crash seemed to stun him.

Stripped almost naked as the man was, his shirt and kilt torn to tatters, and his hose and brogues nearly gone, he still contrived, by means of his garters and shot belt, to secure the deer, by binding his hind leg to a birch tree. Having accomplished this with great difficulty, he returned for his gun, and thus at length secured his victim.

If that vast tract of land in the extreme north, designated as "Lord Reay's Country," has produced some wild and ferocious characters, it has likewise tempered its romantic district by giving birth to a man of no ordinary celebrity. Rob Doun, or brown Robert, was born in the heart of it, at Durness, in the year 1714; and although a distinguished bard in his time, would probably have sunk into oblivion had he not fortunately been rescued from it by a publication of his Poems, and an Essay, prefixed to them, by the Rev. Dr. Mackay, minister of Laggan. Rob could neither write nor read; nor was he much of a philosopher: there were no academic groves in the wild land of his fathers. "But the habits of oral recitation were in vigour all about him," and being, by nature, endowed with a rich fancy, and a retentive memory, his mind was stored with romantic legends and superstitions, which, perhaps, abound more in that district than in any other part of Scotland.

The following account of this northern bard I have extracted from the Edinburgh Review, for July, 1831, with some variation, however, for the sake of compression:—

"His witty sayings, his satires, his elegies, and, above all, his love songs, had begun to make him famous not only in his native glen, but wherever the herdsmen of a thousand hills could carry a stanza or an anecdote. Donald Lord Reay, a true-hearted chief, resident constantly amidst his 'children,' and participating in all their affections, presently claimed for himself the care of the rising bard of Mackay; and Rob was invested with the office of *boman*, or head cattle keeper, an employment which, at that time, carried with it abundance of respect in the eyes of his fellow mountaineers.

"Rob was an inveterate deer-stalker; from earliest youth it had been his delight to spend days, nights, and even weeks among the wildernesses, in pursuit of this spirit-stirring diversion; and, among prouder titles to distinction, his kinsmen honoured him as a marksman of the first order, and a proficient in the mountain chase. In his boyish days no one had ever dreamt of restraining indulgences of this kind; and though now law had been added to law, and regulation to regulation, 'honest theft is the spoil of the wild deer' continued to be a proverb in every mouth, and even the *boman* of Lord Reay was a constant trespasser; often had he narrowly escaped the arm of the law, and yet nothing seemed capable of converting him from his darling error."

"He was more than once," says the writer of his memoirs, "detected in the forbidden act, and in due time summoned before the sheriff-substitute, when, in event of sufficient evidence, the issue must have been banishment to the Colonies, in terms of the statute. An anecdote on this occasion, strongly characteristic of the bard, has been lately related to us by his still surviving daughter. He set out to attend the court early in the morning, attended by a neighbour, one of his wonted hunting companions. The prospect of transportation pressed heavily on his friend's spirit; but the bard remained seemingly quite tranquil. Not so his wife, who, with lamentations and tears, could not be prevented from accompanying her husband a part of the way. The bard would not, even now, part with his favourite rifle, but shouldered it at departing with his wonted glee. 'It was,' said his daughter, in reciting this anecdote in the Gaelic tongue, ' Bha gunna caol, dubh, fada, mallaicht aige,' that is, a slender, black, long, *wicked* gun which he had. They had not proceeded beyond a mile from home when they came full upon a small herd of deer; Rob was not to be restrained. He fired and shot two of them dead upon the spot. His wife, before in extreme consternation, was not now to be pacified. She imagined that her husband had just sealed his doom. He beseeched her to be silent. 'Go home,' said he, 'and send for them; if I return not, you shall have more need for them;' but, saluting her, he added, in kindlier terms, 'fear not, it shall go hard with me if I am not soon with you again to have my share.' The truth was, that, though threatened by the authorities, there was scarcely one of the country gentlemen who would not have gone any length to protect the bard from the violence of the law."

This action, and some satirical ballads written by our bard, created a coolness between Rob Doun and his chief; but he obtained protection afterwards in the family of Colonel Mackay.

I conclude this notice with a short extract from one of his translated songs, written after a long absence from the object of his love, who eventually proved faithless:—"the home-sickness it expresses appears to be almost as much that of the deer-stalker, as of the loving swain."

> "Oh, for the day for turning my face homeward,
> That I may see the maiden of beauty:—
> Joyful will it be to me with thee,
> Fair girl with the long heavy locks!
>
> Choice of all places for deer-hunting
> Are the brindled rock and the ridge!
> How sweet at evening, to be dragging the slain deer
> Downwards along the Piper's Cairn!
>
> Easy is my bed,—it is easy;
> But it is not to sleep that I incline:
> The wind whistles northwards, northwards,
> And my thoughts move with it."

To this account of poachers and freebooters, already I fear too long, I venture only to add a notice of a very singular trial which took place at Edinburgh, on the 10th of June, 1754.

Duncan Terig, alias Clerk, and Alexander Bain Macdonald, both notorious poachers, and reputed freebooters, were indicted at the instance of His Majesty's advocate, for the murder of Arthur Davies, sergeant in General Guise's regiment of foot, in the year 1749. The trial, though not of an unprecedented nature, involves a very curious point of evidence, and was printed in 1831, at the expense of Sir Walter Scott, and presented by him to the members of the Bannatyne Club. Its circulation being thus limited, I am glad of an opportunity of inserting Sir Walter's remarks upon, it, which are probably novel to the majority of the public.

"The cause of this trial," says Sir Walter, "bloody and sad enough in its own nature, was one of the acts of violence which were the natural consequences of the civil war in 1745.

"It was about three years after the battle of Culloden, that this poor man, Sergeant Davies, was quartered with a small military party, in an uncommonly wild part of the Highlands, near the country of the Farquharsons, as it is called, and adjacent to that which is now the property of the Earl of Fife. A more waste tract of mountain and bog, rocks and ravines, extending from Dubrach to Glenshee, without habitations of any kind, until you reach Glen-Clunie, is scarcely to be met with in Scotland. A more fit locality, therefore, for a deed of murder could hardly be pointed out, nor one which could tend more to agitate superstitious feelings. The hill of Christie, on which the murder was actually committed, is a local name, which is probably known in the country, though the Editor has been unable to discover it more specially, but it certainly forms part of the ridge to which the general description applies. Davies was attached to the country where he had his residence, by the great plenty of sport which it afforded; and when dispatched upon duty across these mountains, he usually went at some distance from his men, and followed his game, without regarding the hints thrown out about danger from the country people. To this he was exposed, not only from his being entrusted with the odious office of depriving the people of their arms and national dress, but still more, from his usually carrying about with him a stock of money and valuables, considerable for the time and period, and enough of itself to be a temptation to his murder.

"On the 28th day of September the sergeant set forth, along with a party which was to communicate with a separate party of English soldiers at Glenshee; but when Davies's men came to the place of rendezvous, their commander was not with them, and the privates could only say that they had heard the report of his gun after he had parted from them on his solitary sport. In short, Sergeant Arthur Davies was seen no more in this life, and his remains were long sought for in vain. At length a native of the country, named M'Pherson, made it known to more than one person, that the spirit of the unfortunate huntsman had appeared to him, and told him he had been murdered by two Highlanders, natives of the country, named Duncan Terig, alias Clerk, and Alexander Bain Macdonald. Proofs accumulated; and a person was even found to bear witness, that lying in concealment upon the hill of Christie (the spot where poor Davies was killed), he and another man, now dead, saw the crime committed with their own eyes. A girl, whom Clerk afterwards married, was nearly at the same time seen in possession of two valuable rings, which the sergeant used to have about his person. Lastly, the counsel and agents of the prisoners were convinced of their guilt. Yet, notwithstanding all these suspicious circumstances, the panels were ultimately acquitted by the jury.

"This was chiefly owing to the ridicule thrown upon the story by the incident of the ghost, which was enhanced seemingly, if not in reality, by the ghost-seer stating the spirit to have spoken as good Gaelic as he had ever heard in Lochaber.

"'Pretty well,' answered Mr. Macintosh, 'for the ghost of an English sergeant!' This was, indeed, no sound jest, for there was nothing more ridiculous in a ghost speaking a language which he did not understand when in the body, than there was in his appearing at all. But still the counsel had a right to seize upon whatever could benefit his client; and there is no doubt that this observation rendered the evidence of the spectre yet more ridiculous; in short, it is probable that the ghost of Sergeant Davies, had he actually been to devise how to prevent these two men from being executed for his own murder, could hardly have contrived a better mode than by the apparition in the manner which was sworn to.

"The most rational **supposition** seems **to be, that the crime had come to** M'Pherson's (the ghost-seer) **knowledge,** by ordinary means, of **which there is some evidence; but desiring** to have a reason for communicating it, which could **not be** objected to by the people of the country, he had invented this machinery of the ghost, whose commands, according to Highland belief, were not to be disobeyed. If such were his motives, his legend, though it seemed to set his own tongue **at** liberty upon the subject, yet impressed on his evidence **the fate** of Cassandra's prophecies, that, however **true,** it **should not** have **the fortune to be** believed." *

CHAPTER IX.

Broad Awake.—Arrangements for the Day.—A Ticklish Point.—Serpentine Movements.—Disappointment.—White Kid Gloves—Contest of Skill.—Escape of the Deer.—Good Sport.—Close Combat.—A Ride on a Stag.—Remarkable Prowess.—Contest with a Phoca.—The Drive Begins.—Shots and Untoward Accident.—Corrie's Sagacity and Night-Watch.—The Coup d'Essai.—Past Deeds.—Eagles Killed by a Boy.—Driving the Herd.—Legend of Fraser's Cairn.—The Lord of Lovat's Raid.—Strong Taint of Deer.—Nervous Excitement.—Ambuscade at the Wood.—Noble Sport.—The **Old Blair Pony.—Return to the Castle.**

" What is a gentleman without his recreations?"
Cornish Comedy.

"JAMIESON desires **me to tell you, sir, that there are three** fine harts feeding **on the swell of Ben Derig, high above** the cottage, and he **thinks you had better get up, for it is** five o'clock."

"**A** goodly warning, John; make ready our **breakfast** immediately, and let the hill-men swallow theirs as quickly **as possible.** I will call Mr. Lightfoot myself,"

"What ho! hillo, hillo, comrade! **Up, up, and be stir-**ring!"

---

* The trial **of these** men is curious and interesting, but too long for insertion in these pages. I have, however, ventured to copy out the evidence of the two ghost-seers, which contains the chief points in it, and to insert them in an appendix.

"Eh!—what—where—when? comest thou to draw Priam's curtains in the dead of night?"

"Night! now by him who sits on high Olympus,—

'Night's candles are burnt out, and jocund morn
Sits tiptoe on the misty mountain's top.'

There are twenty harts cropping the heather bells on the Red Mountain, just above the lodge.

'Falsely luxurious, will not man awake?'"

"Oh, as for that, you see, I'm quite alive;—yaw—yaw!—confoundedly stiff though: I do not think that arquebusade of yours is genuine. But you'll give me time to put on my clothes, won't you? and although you dispense with sleep, I pray you not to dispense with breakfast. I always eat before I go out; my father and mother did so before me. Now here I am, you see, fresh as a lark; just give me a helping hand will you, my good friend? Thank you: now then, on goes my best jacket; for this day I mean to do 'a deed of mighty note.'"

"Bravo! up with you, then, my good fellow, *quanto primâ*. In the meantime, I will go out and examine the three harts."

"Three!—THREE harts! why, thou said'st twenty ere now!"

"Aye, in buckram: twenty deer, you know, will rouse your somnolent man sooner than three: there is a great charm in numbers."

Tortoise clapped on his bonnet, and marched forth with his telescope, all unclad as he was, save in slippers and dressing-gown. The harts were perused, and found prodigious, of course. A brief toilette—a breakfast short, sharp, and decisive, and perhaps a cauker, as the Ettrick Shepherd has it.

All now were ready and about to start, when a hill-man came panting in with a letter from the Duke of Atholl containing instructions for Tortoise to bring down as many deer as he could, and to be at the Green Knowes at three o'clock, mentioning where he and his parties would be posted, and saying that his men would keep them in on the west after they had passed a certain point. In the meantime, Tortoise and his friend were to kill what they could.

This was pleasant news. They had a long day before them, and plenty of time for all operations, both private and public.

Now if truth were told, the harts above mentioned were on the round even swell of the hill, where it was judged very difficult, if not impossible, to come within distance of them: " But say nothing of this, Jamieson, we must do our best. We will not throw a damp over the chance."

A stony burn comes down from the mountains near Bruar Lodge, which has hollowed out a deep chasm between two hills. The eye of no living thing can command this narrow pass from the heights above. Up this water-course the party proceeded, over fragments of rock, through the streams, and little linns, directing their steps towards the east, it being judged best to endeavour to come in by a side wind from that quarter. They continued to ascend the burn for a long time, happy when the disposition of the ground permitted them to step out for a space on the heather. At length they gained the ascent, and from a black bog, which they had entered, discovered with their glasses that the deer were still in the same situation. A death-like silence took place: the ground was examined minutely. Then the glasses were closed, and deep thought and care sat on the countenances of the sportsmen. The bog which had hitherto been their cover, terminated long, long before they could get within any reasonable distance of the deer, who were, moreover, in a commanding situation. The men had observed a ridge of high heather, insufficient, they judged, to conceal them; that, however, must be tried as their only chance: the dogs were left in the bog. Lightfoot's rifle was given to Jamieson, and they crept cautiously out of the hole, where they had been skulking. Their caps they put in their pockets, and began to writhe themselves through the heather like serpents. The ground was dry, but the operation was tedious, and even painful, so that they took occasional moments of rest. They dared not raise their heads ever so little out of the dewy heather, which they shaved so closely that there was scarcely a waistcoat button left in the party. They strove with their feet, and clawed with their hands, still making but slow progress.

At length their hearts throbbed with nervous excitement, for they were fairly within a hundred yards of a long shot. For a space they rested to ease their limbs and gain steadiness, still lying extended like corpses. Tortoise whispered, "Now then be calm, and when we come within distance, take the hart to the right,—he is the best; a little further and our task is done."

Twenty yards forwarder they gained in security; another ten with the same success;—they were getting nearer and nearer every moment, and their hearts trembled. There was a little knoll, or small rise of ground, before them, where the heather grew in larger tufts, and this point once gained (of which there was every probability), they would be within reasonable distance of as fine harts, they roundly asserted, as any in the forest; so onward they still crawled, with pain and fatigue.

But if deer-stalking, or any other species of sporting, were of easy achievement, what would become of all those delightful changes that animate us in the chase? no longer would our bosoms throb with hope, or sink from an apprehension of failure; we should keep "the even tenor of our way," tame in pursuit of the quarry; and, as Captain Bobadil has it, "too respectful of nature's fair lineaments." Plans well laid and executed,—difficulties overcome by skill, by labour, and perseverance,—these are the events that flatter our self-complacency, and give the highest zest to the sportsman.

It is the desire to evince this skill, and surmount these difficulties, that carries the ardent deer-stalker through bog, through burn, up hill, and down precipice; creeping, wading, running, or lying; heedless alike of mire, waters, and fatigue: but still with all his caution, even with the most consummate generalship, and in the very tumult of expected success,—

" ⸺ medio de fonte leporum,
Surgit amari aliquid quod in ipsis floribus angat."

And if ever a bitter thing did happen, if ever the chalice were dashed from the lips, it was at the critical moment when we left our sportsmen just within shot of the deer.

"Tears of compassion tremble on our eyelids," whilst we

are obliged to recount, that an old chuckling moorcock sprung from those very bunches of heather, which they vainly thought their haven.

Oh Puck! Puck! why didst thou place that officious bird in that particular spot, to scare away the deer? was there no other place in all this wide forest where he could set his breast? A thousand, ten thousand there are, where surely he might have been as happy; it was a chance as one to a million: see what a pickle we are in; mark what we have done, what endured! But thou delightest in mischief, and art grinning, I know, thou impious little elf, and, *maledetto che tu sia*, wert never better pleased in all thy life. The deer, thus warned, broke over the hill, and the moor-cock went darting away, turning himself side-ways to catch the gale with his wing, chuckling, and rejoicing, as it were, in his free flight and the success of his mischief. "Now may a dart from Murdoch's quiver pierce thy side before night!"

"Well, it was not our fault, that is some comfort, there was no kid glove in the matter; an allusion you will better understand, when I tell you that a celebrated sportsman, after having made a very long and laborious circuit to come into a *quiet shot*, destroyed his chance, when on the very verge of attaining it, by a slight elevation of one of his hands which was decked with a white kid glove: it is marvellous how such a piece of furniture found its way into a Scotch forest; and one is tempted to exclaim, in the words of Mrs. Siddons,—'*How gat it there?*'" [*]

The sportsmen arose, and put the best countenance they could upon the matter, which, sooth to say, was no better than a very doleful one, deadened as their hearts were by disappointment. The deer, however, had not seen them, and were still in the ground before them. In fact, when they came over the hill, they saw them looking back jealously in the moss below.

---

[*] I do not vouch for the tale, but it is said that Mrs. Siddons, hearing a story about a French official who was locked up in his *bureau*, being rather in an absent mood, fancied that he had been thrust into a chest of drawers, and exclaimed, with great pathos, "*Poor gentleman! how gat he there?*"

"There is no coming down upon them from the hill," said Tortoise.

"They will no bide there lang," said Jamieson.

"They are magnificent creatures," said Lightfoot.

"Shall I lowse a doug?" said Maclaren.

"No, that may spoil the drive, for there is no saying where a cold hart may go to bay : but stay you here ; we will take a long round, and endeavour to get into the burn. Give us twenty minutes, and then try to coax them across, as near yon curve of the stream as you can. If they move forward, we will do so too ; so keep the glass upon us, and do your best according to circumstances.—Now look at your watch."

A long round, and a sharp persevering pace, brought them to the destined spot within the allotted time : having walked for a space with bent bodies, they sat themselves down on a grey stone under the bank of the stream. Maclaren now began his game ; entertaining enough it was to see the contest of skill between him and the harts : the continual shiftings of the Gael, however, at length gave them a slight turn towards the east, and they appeared to be coming in a good accommodating direction. But whether they got a blink of the men in the burn, or found ground more to their liking, they at length kept full to the wind, and went straight south. The moss-troopers had not as yet been able to come forward on account of the wind ; but now that the course of the deer was obviously determined upon, they made the best of their way under cover of the banks and bogs. All too late they were ; for the harts crossed the burn out of distance, but at a slow pace, as they saw not the men.

Maclaren now got as well round to the west as time would permit him to do : but it was not this manœuvre that made them again bear a point to the east, for they held him particularly cheap ; it was rather the sight of a few hinds that had been disturbed from under the crescent of Ben-Dairg, and were bearing away towards Cairn-cherie.— These they meant to join.

"This way, this way, Harry, come along, we'll have them

yet." So saying they strove through the deep channels of the peat moss, cowering low, and cutting off the angle with all speed, till they got fairly within shot.

Now there is one point in deer-stalking that is the most provoking and tantalising thing in the world; and as it happens pretty often, so it occurred in the present instance. The riflemen, I said, were within distance: so indeed they were; but the harts skulked up a deep channel in the moss in such a way that nothing but the points of their horns were to be seen. It was in vain to run after so small a parcel; that would only give them a rapid start, and set them clean away at once. Thus not a shot was fired, and fortunately no one committed suicide.

The harts now joined the hinds, and all went slowly up the western face of Cairn-cherie.

As soon as they were all fairly settled in their new position, a fresh reconnoissance took place—the deer had so taken up their ground that they were not within shot either from the top or base of the hill, each of which points the riflemen could have gained unseen by them. The party went forward to the foot of the hill.

"Noble fellows! cunning devils! what is to be done now?" said Lightfoot, "Do you think we shall kill them all? Can you bring us near them in fine style; cannot we come down upon them from the rocks above, and put them all to sudden death?"

"To sudden flight we may easily: but know that there is no place in the whole forest so ticklish as this; the ground, on the summit, is so varied with high rocks, flats, and hollows, that currents and swells of air pass in almost all directions, and the difficulty will be to get near the deer, and keep the wind; but I know the ground well; aye, every inch of it, quite as well as my own cabin at Bruar."

"Aye, ye're weel acquent with it, for the beasties ha' bin ow'r canny for ye whiles amang thae rocks."

"Hush, hush, my good fellow, no tales."

"Na, I canna but say that ye ha' had good sport there too, but sure ye'll no be forgettin the big hart that gat a gliff o' ye, and skelped awa through the moss, joost as ye war thinkin to pit yer ball intill him; perhaps Mr. Lightfoot would like to hear something anent it?"

"No, no, Maclaren, I know he would not; let by-ganes be by-ganes. So now tell me, what is your advice?"

"Why, I wud ha' ye advised to gang round to the east, and to leave me at the fut o' the hill; ye can win to the tap in ten minutes, and when ye are there, I can pit ower the deer. But ye mun be canny, and ye mun aye throw out wee bits of tow, for the wind is unco kittle among the rocks; ye'll bear in mind the muckle hart—him that ran awa sae brawly frae ye, without skaith, when yer honour thoucht to hae takkin his gralloch, and said something anent his tallow, and white puddins, and the fat on his haunches."

"You advise well, Maclaren, and your discourse is voluble; sweet I may not say, since the latter part of it falls somewhat unseemly on my ear. Now look at your watch, give us a quarter of an hour; start the deer to the moment, as quietly as possible, they will be ready enough to come, without compulsion."

"So here is another of your boggy steeps,—antiseptic no doubt; but I will not be buried in them to try their properties: I shall get up capitally."

"Not if you proceed in that manner, I assure you. This hill is too steep to walk heel and toe; your style is not mechanical: see what a lever you are making use of; just stick the side of one foot horizontally against the hill, and bring up your other underneath it, keeping the same foot always uppermost as I do: see now how compact you go without labour, almost without exertion, and certainly without the aid of your hands, which you were using before."

"Capital! so I do. Can you also give me any receipt for running?"

"Only, as I said before, to go as compact as possible; all swinging of the arms, and kicking of the legs behind, is so much unnecessary motion, which impedes your progress, worries the whole body, and distresses your wind. But a truce to conversation, however agreeable to me: we must now proceed in silence."

Now had they passed the moss, and attained the rocks on the summit, and were sitting down behind a large block of granite; they laid the rifles on the ground, pulled off their caps, and wiped their foreheads—Tortoise held his watch in his hand; it wanted five minutes of the time for starting the deer. Again and again he looked at the slow progress of the minute-hand: it was just on the point; it has passed it; the deer then must be in motion: a short space he gave them, to get forward, that he might be secure of the wind; then, snatching up a rifle in one hand, Jamieson following him with another, he waved his hand to Lightfoot, got quickly forward, and clambered up a rock, where all posted themselves aloft.

They had not been in this commanding position five seconds before the deer came racing below them over the naked ground, at an easy distance. When Lightfoot saw the hinds, who were leading, he was in the act of raising his rifle, but his arm was immediately arrested by Tortoise, who continued to hold it with a significant look, but in silence. Now came one of the wonderful harts; he was a stupendous animal, very sleek in his coat, and had royal antlers; that is to say, three points on each horn. "Take him," said Tortoise, letting go his friend's arm; "and fire well forward."

The old rocks of Cairn-cherie rang to the rifle sound; the deer slackened his pace, and then stood still. This shot had scarcely been fired, before another monster came in view. Tortoise levelled; the fatal trigger was pulled,— the hart catched his side, stood for a space, and then went slowly on with the rest. The third hart swerved a little below the hill, and never came within distance.

"Joy to you, my friend, your deer is safe enough, and so is mine, I hope. Lie still, for Heaven's sake, or you will spoil all; he is sick—he is dying!"

The poor fellow stood for a short space, with his forelegs extended; his knees then bent a little; his head rose and fell alternately for a few moments; his whole frame quivered, and down he sank to eternal rest. The pangs of death were brief, but very painful to witness. They now went forward, and the knife was plunged into him, when his blood gushed out in torrents. A man was left to gralloch him. Lightfoot could not be torn away from the dun beauty: the hill-man, as he gralloched the deer, and drank the whiskey, swore there never was such a deer seen in the forest; he grew larger and larger at every quaigh-full, and there was no saying to what a portentous size he might have arrived, had not the flask been fairly drunk out.

The rest of the party went slowly forward, till at length they saw the other wounded hart lying in a bog. He was extended, and kept his head as low as possible; it was apparent, then, that he was not only alive, but had his senses about him. Tortoise crept cautiously up, and sent a ball through the back of his head—as deadly a shot as can be made.

The smile of joy danced on every countenance, but chiefly on thine, O Lightfoot; the warm current came tingling through your veins: there was a buoyancy of spirit, and an air of success about you that proclaimed you a king—a hero—a demigod! Hercules was a pretty fellow; so was Theseus; so was Pirithous; but, although they subdued various monsters, they probably never killed so fine a stag in all their lives. Happy, thrice happy mortal! happier far than Candide, when he met Miss Cunègonde amongst the Turks, or (to make a more apt comparison) than our own Phidias,* when he killed two woodcocks at one shot. Thou shouldst have died that moment, my own hero: alas, why did you survive, to pace over geometrical enclosures in pursuit of pigmy game? But bear thy faculties meekly, whilst the deer are being gralloched, and the black flag is hung on the bonny antler to scare away the raven.

"Now, Tortoise, I really think that Macrobius, and the rest of Virgil's commentators, are senseless goupies; for I am ready to maintain, in spite of them all, that the slaying of such a magnificent animal as this was a very fit cause for the Latin war—

'Cervus erat formâ præstanti, et cornibus ingens.'

---

* Who has not heard of Sir Francis Chantrey's skill with his gun and his fishing-rod? The above incident occurred at a great battue—at the Duke of Bedford's probably—and the whole party saluted Sir Francis on the occasion with solemn deference, each individual passing before him in succession, and making his obeisance.

(Handling the horns all the while.) But why did you throw away a charge upon your wounded deer, who was lying extended in the bog, and at your mercy? I should have preferred close combat, like our friend the artist; I would have got across him, and seized him by the horns."

"In which case you would have had a charming ride, like the late Glengarry, or like the forester of the present chief of Clanchattan, who, in passing last summer* through the forest of Stramashie, near Loch Laggan, descried the horns of a stag above the heather at some distance; and taking advantage of the cover of a grey stone on the lee-side of the animal's lair, crept cautiously up to him, whilst he was apparently asleep. He had no rifle, but opened his deer-knife, which he placed between his teeth that his hands might be free, and then threw himself suddenly upon the stag. Up started the astonished beast, and sprung forward with Donald on his back, who grasped him with might and main by the horns, to keep his seat in a sportsmanlike manner. No easy matter, I trow, for the animal made right down the rugged side of a hill with headlong speed, to a stream in the glen below, and dashed through it, still bearing his anxious rider with the knife in his mouth, which he had neither time nor ability to use. When, however, this gallant pair reached the opposite side of the glen, and the deer began to breast the hill and relax his speed, Donald was enabled so far to collect his bewildered senses as to get hold of his knife; and he absolutely contrived to plunge it into his throat. The deer fell forward in the death-struggle, and Donald made a summerset of course. In consequence of this extraordinary feat, the man has been dubbed by the people with a new and appropriate name in Gaelic, which my authority (Mr. Skene) told me he could not pretend either to write or to pronounce. This was dexterous work; but there are innumerable examples of the spirit and determination of Scottish sportsmen: and whilst the deer are being gralloched, I may as well relate an adventure that happened to a celebrated and enthusiastic deer-stalker, whose name I am not at liberty to mention.

---

* The summer of 1837.

"Whilst hunting lately in the island of Jura with his deer-hounds (for he seldom carried a rifle), he came rather suddenly upon three magnificent stags: he slipped his three dogs upon them, and what is very singular, and proves their spirit, each of them took a separate deer, and they all went in different directions. After a long and arduous pursuit over the rough hills of Jura, the stalker* at length got sight of one of the deer standing at bay in some long heather, in a deep hollow: he appeared to be quite exhausted; and the dog Oscar, one of the most powerful and intrepid of the breed, was lying within a few yards of him apparently done out. As soon, however, as his master shouted his name, the gallant brute sprang at the stag's throat, and a desperate battle ensued, in which the dog was tossed three times in the air before his owner could get quite up, and was thus severely wounded.

"When the sportsman, who had only a little herd-boy with him, reached the arena, the stag, without attempting to make off, thrust at them right and left, whirling round and round to defeat every attempt to grapple with him; the boy had his leg severely lacerated, when the deer-stalker, who is a most muscular and powerful man, dashed in, and seized the animal by the horns. The contest was desperate and doubtful; at length they both came to the ground, when the hunting-knife finished the contest.

"This same gentleman, whilst shooting sea-fowl, amongst the rocks of Colonsay, perceived a large seal basking on the shore; he drew cautiously towards the spot, and gave him the contents of his fowling piece, when the seal scuffled over the rocks, in his way to his element. Our enthusiastic sportsman sprung from the boat, and, grappling with the slippery brute just as he had reached the water, plunged headlong with him into the sea, where a singular conflict ensued, sometimes under water and sometimes in view, before the people in the boat could manage to get hold of either of the combatants; at length, however, they succeeded in dragging both the young laird and his fat friend into the boat, to the great merriment and relief of his

---

* It is necessary to be a good stalker in order to lay on the dogs properly.

companions,—to whose remonstrances he only answered, 'D—n the brute! Did he think to give me the go-by?'

"Mr. Skene, who told me this anecdote, was himself the prototype of Sir Walter Scott's story of Highland Hector's contest with the phoca, in the *Antiquary;* having related to him on the spot an encounter which he had with seals in descending the rocks at Dunotter, in his passage to a creek, from whence he proposed to make a sketch of the castle."

During this relation the hill-man stripped off his grey jacket, bared his sinewy arm, and went through the necessary operations of bleeding and gralloching. Every movement, every finesse was exultingly run over;—the dogs fought; the men laughed and drank; and were as cordial as success and right good Loch Rannoch could make them.

"But the day wears apace; we must now separate our forces, and if we forget not our cunning, we will sweep these glens and mountains, and put down such an army of deer as shall give free exercise to the rifles from Blair; their volleying shall scare the roe in his secret glade, and visions of the magnificent herd shall again warm the imagination of the Southron in his festive halls, and great shall be the boast of those who were present on St. Crispin's Day."

"Heyday! Why you affect to be Ossianic to-day! And, upon my word, what with the mountain air and scenery, and the heroic deed I have just done, I tread the heather with something of the feeling of a descendant of Fingal myself. But, allons, cater we now for the general sport; and here shall end our stalking; here on the old rocks of Cairn-cherie, never to be forgotten, till we depart to where Tullus and Ancus have gone before us."

The party now began to occupy their posts. The riflemen remained on the middle hill; Maclaren was sent across by the Craggan-Breach to Sroin-a-chro, and Sandy Macintosh to Ben-y-chait. All came forward at the signal, which was the exposure of some man's shirt, by means of unbuttoning his waistcoat; a luminous mark, that could be readily discerned through the telescope, which each man carried with him, placed in a leathern case and slung in a belt across his shoulders.

The sport now about to take place, as far as driving went, was very similar to that practised in a deer-drive to Glen Tilt; but in the termination it differed materially; for instead of running the gauntlet as the deer did at Glen Tilt, and passing freely onward to the heights of Ben-y-gloe, they were, in this instance, to be pressed on to the pine wood, that formed the barrier between the mountain slope and the cultivated strath of the Tay. This wood was held by them a place of refuge; and when they gained it the sport was understood to be terminated, though a hart or two might occasionally be killed after their entrance into it. These woods are fenced on the moor-side by a stone dike, and behind this dike some of the parties that came from Blair were posted; so that the little army of deer were thus placed between two fires—that is to say, between the rifles of the sportsmen who brought them down from the mountains, and those who opposed their passage into the wood: thus beset, in front and rear, and at their flanks, all their sagacity was called forth; and their movements being more varied, were by so much the more interesting. The difficult point was, for those who placed themselves in front of the driven deer, to avoid giving them their wind prematurely, which might be managed by keeping at first to the east and west (the wind being south), and drawing towards the centre when time served.

Let us now see what the hill-men were about.

After a lapse of about forty minutes the men had arrived at the stations above mentioned, and the signal was given for starting. There were groups of deer both in Glen Mark and Glen Dirie—hinds, calves, and a few harts: very little management was required to get these forward, as they naturally, and readily, went up wind; which was all that was required of them. So they were urged forward, and driven out of the glens, with shouting and hurling of stones, which bounded down the precipices with repeated echo to the vast depths below. Still, as the men came onward, the deer joined their forces, formed, looked back calmly, and, as usual, scrutinised every part of the ground on their flanks, and on their rear. Tortoise had given up all thoughts of manœuvring any more for himself and his friend; but as he did not seek sport, so he was determined not to shun it if it were thrust upon him. And fortune (who seldom does things by halves) now placed another of her favours in his way. Whether or not he benefited by the chance will be seen in the sequel.

Thus then it was: a few hinds and calves, with a good hart amongst them, came rapidly over the shank of the hill which he and Lightfoot were descending. The hart was generally masked by the hinds; but as their paces were unequal, he was sometimes exposed for a moment. Both sportsmen suddenly clapped up the rifles to their shoulders: the point was too nice a one for ceremony. The fatal sound of Lightfoot's was first borne along the moor—fatal did I say? fatal to what? Alas, to the hind that was coming up in the rear of the hart; down she dropped, and her maternal cares ceased for ever. In the meantime Tortoise kept holding pertinaciously where the hart was, keeping his gun well forward: half of him at length was clear, the trigger was instantly pulled, and the ball took effect; but the wounded stag went on behind the others, and the men couched down upon the heather blossom.

And now happened one of those untoward accidents that will sometimes occur in spite of ordinary precaution. The dogs had been brought forward for the stricken deer; and Corrie, who had a small greyhound-like head, slipped himself from the leash, and away he went on the traces of the deer. Nothing could be more agonising, for there was every probability that he would put the main herd out of the cast, and disappoint all the parties at the wood. But come what might, the keen hound was gone forth, and no earthly power could arrest him.

The small parcel that had been fired at joined the great herd, full in Corrie's view; and all disappeared for a while in the hollow of a deep ravine, with the dog at their traces. But they soon reappeared on the opposite brae, Corrie being still close upon them: every man was absolutely in despair. He forced them into a compact mass, ran furiously at their rear, then to one flank and then to the other; and ever as he came on, the outward deer endeavoured to wedge themselves into the mass out of reach of his horrid fangs.

There was now no doubt but that the drive would be spoiled. Many were the denunciations against the appalled leashman; his death-warrant was made out, for he was to have no more whiskey, which was precisely the same thing to him.

But, lo! when all were sinking with apprehension, affairs took an almost miraculous turn: after the hound had forced the herd in the manner described, missing the taint of the blood, he suddenly turned back from them, and came feathering along, making beautiful casts to the right and left; returning now to the burn which he had before passed, he picked up the lost scent of the blood, and ran rapidly down its mazes. Soon the wounded deer sprang up, and went heavily before him down the stream; out at once leaped the cunning dog upon the banks, headed him about a hundred yards, and then came back in his front, and held him resolutely to bay. It was a way he had of shortening the business.

This happy termination was an inexpressible relief to all. Tortoise went forward alone, creeping up cautiously by a side-wind, and finished him by a shot through the head. When the men returned to the hind, they saw the eagle sweep down from the clouds, and wheeling over Ben-y-venie, descend in all his expanse of wing, and perch himself upon the blasted branch of a birch stump that overhung a rock in the declivity. There the huge bird sat the whole time the deer was being cleaned, gloating over the operations, and eager for the bloody repast. As soon as the animal should be left on the lonely moor, he thought to cower over him, uttering his shrill shrieks, and to plunge his beak into the eyes, and pick them from their sockets. But the foul bird shall be baulked of his prey. The sagacious Corrie shall protect him; Corrie, who will never leave a dead deer without compulsion, but will coil himself up by his side, and watch by him during the chill blasts of a northern night, guarding him till the hill-man comes in the morning to cord him on his sheltie; then the good dog will once more lick over his dun sides, shake his tail, and fawn upon the hill-man, and escort him home to the slaughter-house. Corrie would do all this as well as the rest of his litter; nay, if he were slipped on the moor, he would go back alone to the last deer that was killed, although it were many miles distant, and protect it through the night from the fox, the wild cat, the eagle, or the raven.*

All now good-humouredly tried to make out the hind a yeld one; but it would not do; she evidently gave suck, and was also singularly lean.

"Never mind, Lightfoot; she richly deserved her fate; for it was a wicked deed to place herself where she did. So pray be comforted."

"No, no, it will not do. The Badenoch fairy's speech rings in my ears, saying, or seeming to say, 'O Lightfoot, Lightfoot, thou hast this day slain the only maid in Doune.'"

"Never mind, these things occur to us all; the hart had a very narrow escape from your ball. You heard our friend from the south brag the other day how nearly he had killed a deer; and when you asked him in what manner, he replied that his ball struck the spot where the deer had been lying the day before. You were much nearer than this, you know. It was no bad shot after all, and will be of infinite service to you as an instruction to take your aim forwarder in future. I began my career nearly in the same way, and learned a good lesson from it."

"Then the first deer you killed was a hind? Well, that's some comfort, however."

"No, I mistake; not the first. My *coup d'essai* was at a hart. I set off from Blair Castle with the Duke of Atholl for Forest Lodge at twenty minutes past three o'clock in the morning. There were no deer feeding in the glen; so we breakfasted, and I began fishing for salmon. After a time, whilst very intent on my cast, I heard a noise above me, and, looking round, I saw a stag running at full speed along the slope of the hill, with two lurchers at his heels. Quickly did I clamber up the rocks. John Crerer was in the road with a rifle; and, as he was in the act of raising it to his shoulder, in I came behind, took it from his hand, fired, and hit the deer through the jaw. The poor chop-fallen fellow then went to bay, where I finished him; but,

---

* A beautiful painting, by Mr. Edwin Landseer, of this sagacious dog, thus engaged, will be in the recollection of many.

to speak the truth, he was altogether as lean, ragged, and shabby a beast as I ever saw. If I was not ashamed of him, I am a soused gurnet."

(Maclaren, touching his hat.) "Ye held at better game afterwards atween the shank of Ben-y-chait and the Elrich, when Charlie Crerer was with ye. Ye'll mind when ye creepit up to four harts to tak' a quiet shot; ye got within a lang distance, and took the first deer with his braidside towards ye as he was feeding, and lying as ye were yoursel' all alang in the heather, and the ball passed through his heart. And then ye jumpit up, and kilt other twa, ane after the ither, as they were skelping awa', and thus we got three beasties out of four. They say ye steppit the ground afterwards, and that the first deer stuid one hunder and forty yards frae ye. The last must have been an awfu' distance."

"Aye, Peter; a true bill that. More by token that my fingers tingle yet with recollection of the hearty Highland grip that Charlie gave me when he saw the deed; for he's a fine shot, and a dear lover of the sport himself. But if we boast thus of our past deeds, we shall be thought to have lost all hope of equalling them in future."

While thus speaking, Tortoise had been watching the villain eagle. How easily, thought he, I could stop thy murderous career for ever! "Now, Jamieson, could I come in upon that beastie by sinking the hill, going round by the west, and coming up the hollow by a side-wind, whilst his keen eye is fixed upon you and the deer; but the day is far on, and we must be true to our time, and yet it grieves me, for these eagles are very difficult of approach, even by the most skilful sportsman, and it is very seldom one has such a good opportunity. Instances of success, however, sometimes occur; and the most extraordinary one I ever heard of was related to me by my friend Mr. Skene of Rubislaw. Listen to it, Harry.

"Whilst staying with his relation at Abergeldie, he met a herd-boy coming down the avenue, labouring under the burthen of what appeared to be some weighty animal, trailing on the ground behind him, and held by a leg over each shoulder; he concluded it was a roe-deer, but found on coming up that the boy (who was only thirteen years old) had got two magnificent eagles, which he held by the necks over his shoulders, and seemed ready to drop from fatigue.

"It appeared that young Donald's indignation had been roused by having failed a few days before in his attempts to defend a lamb which was carried off in spite of him; and many others of his flock had shared the same fate. Meditating mortal revenge, he got possession of his father's gun by stealth; and marking the eagles to their eyry, in Lachnagan, he hid himself on a rock near the nest, and remained there all night.

"At break of day the male eagle kept hovering about the nest, and the boy took a deliberate aim, and brought him to the ground. The female soared aloft, and stooped after her mate for some time, but out of distance from the boy, who, from fear, dared not venture from his hiding-place, as his prey still struggled amongst the stones at some little distance from him; at length the female eagle flew off, but soon returned with a lamb in her talons for the supply of her young brood. In the meanwhile the determined little rogue had reloaded, and watching his time warily, took another shot, and with such skill and effect that the female fell prostrate and quivering beside her mate; but the poor lamb was killed. Mr. Skene added that he measured the birds at the time, but has mislaid the note of the measure; he well remembers, however, Abergeldie's observation, that they were the largest birds he had ever seen; and most noble animals they certainly were."

The whole herd of deer were now belling, and going lazily up Cairn-dairg-mor; and there they stopped, crowning the hill, and looming large on the sky line. In such vast numbers had they collected, that you might have fancied yourself with Vaillant in the great hunting-grounds of Africa.

The hill-man to the west had shifted his position much farther to that quarter; and the men were so disposed that the deer were kept on the middle hill in a straight line with Blair, with the stalkers in their rear. Thus all promised well hitherto. Tedious it would be to recount the shiftings of the men, which kept the deer in the right course. They were all similar to each other, and the process was a very simple one. When the herd attempted to swerve from the desired direction, the men, who were far distant on the opposite hills, had little else to do than to show themselves in a line, so as to oppose their passage, dodging with them, and taking care not to hurry or press upon them rashly. Had they come too near, the herd would have swept past them in a moment.

"We must now keep back," said Tortoise, "for the deer are examining the ground on the west, and are in no hurry to advance. During this slow operation, I may as well give you the history of Fraser's Cairn, which we passed the other night, when Peter was so valiant about the laird's ghost.

"Tradition informs us that Lord Fraser of Lovat made a raid into the Atholl country, and harried it on his return. This raid was of so ruthless a character, that it was probably executed in revenge for a similar irruption made by the Atholl men on his own demesnes. On the Lord of Lovat's return with his plunder, one Donald Fraser, a clansman who had acted a conspicuous part in the whole business, asked the lord if he did not swear, before going out, that he would leave neither horse, cow, sheep, or cattle, or even cocks and hens, in the Atholl country. 'Ye hae done brawly,' said he, 'and muckle gear hae we gotten; but yon cock that I heard crowing in the toun below us seems to say that the aith is no that completely kept.'

"Lord Lovat demanded if it were a dunghill cock that he heard, or a muir fowl; and upon hearing that it was the former, he replied, 'This must not be; it is against the aith I made ere I set out: get thee doun to the toun, Donald, with a party, and put the beastie to death.'

"Donald did as he was commanded; but upon his arrival, the Atholl men, having had time to assemble, attacked his party, and all were soon slaughtered, except Donald Fraser himself, who was a powerful man, and fought lustily. He was, however, shortly overpowered by numbers; and they proceeded to bind his hands behind his back, that they might make use of him as a guide to conduct them to the spot where the Lord of Lovat was awaiting the return of his men.

"Donald, however, by a sudden and violent exertion contrived to extricate himself from their clutches, and to get a start over the moor; but being encumbered with the cords, which were still about him, was almost instantly overtaken and slain.

"A party of the Atholl men then clad themselves in the tartans of the men they had killed; and, easily making out the track (for the day was now dawning), followed their invaders in a right line, whilst their chief force was kept out of sight in the rear. They soon discovered the Frasers on a swell of the moor before them, but not on the highest point of the ground. They seemed to be regaling themselves with their booty, whilst their horses were grazing around them.

"The Atholl men now sent their main force to the westward by the river Bruar, with instructions for them to come over the hill in the rear of their foes, and fall upon them at a concerted signal. The smaller party, exactly similar in number to those that the Lord of Lovat had sent forth to kill the cock, clad in their tartans, were mistaken for his own men, till of a sudden the wild whoop and whistle peculiar to the clan in their onsets discovered the fatal truth. The foes came upon them at once in their front and rear, and a hot conflict ensued. The Lord of Lovat, who was a heavy man, was slain whilst calling for his horse. Very few escaped the slaughter, and the Atholl men returned victorious with the reclaimed booty. The Frasers were buried on the spot where the cairn now stands which bears their name; and the country people, who dare approach it in the dead of night, assert that they often hear the spirit of Lord Lovat calling for his horse—his horse!"

The deer were now urged on in beautiful style from the Beg of Cairn Dairg. It was like the passage of a little army as their files drew on; some were lost in the hollows—re-appearing, and again sinking out of sight amidst the mazes of the moor. Nothing could be more picturesque than their undulating course;—nothing more gratifying than to reckon the horns marked firmly on the sky line as they passed over the summits.

One hart there was amongst the rest that might be known

from a million. His horns were very white, and his body had a tendency to mouse colour;—sleek and dainty he was all over. It was the third hart which had escaped the rifles on Cairn Cherie.

Stop, caitiff, traitor!—but you may fall yet—

> "Nescis, heu perdite, nescis
> Quem fugias; hostes incurris dum fugis hostem."

It is now the appointed time when the parties were expected in the wood. It was ascertained by their glasses that the Duke's men were properly stationed on Crag Urrard; the drivers therefore continued to get forward the herd, which had collected and rested awhile; now they crowned the Scalp of Meal-Remahr, and went streaming down into the vast basin of Corrie-crombie. Many there were who remained on the hill as sentinels; these, however, joined the rest as Tortoise came on. Maclaren, who was on the east, had been strengthened by a force judiciously placed by John Crerer; and the craft now devolved upon these men. Tortoise and his friend, not daring to come forward, lay down on the heather stumps, conversing in scarcely audible whispers.

"They will pass over Na-Shean-Tulichean, or the green knowes which you see before you: how easily could we have them by getting a little forward! But it must not be; here we will abide; only this: when the great herd have fairly passed over the knowes, should some fatigued beast bring up the rear, 'to stop too fearful, and too *fat* to go,' we shall do no mischief if we get on and salute him with our rifles."

"Hist, hist! by heavens, they are coming! how strong they smell!* They must be very near; I hear their trampling. Heaven bless you, keep down! low—low: do not peep; you will ruin us for ever. Your mouth in the heather, if you please:—close—close; even unto suffoca-

---

* A large herd of deer may be smelt at a very considerable distance, particularly after they have been much driven. The writer of these pages has often been governed in his movements by their taint, when they have been below him amongst the steep crags, over which he could not descend to look, for fear of not being able to recover his ground in time, and thus losing the command of the hill. The taint, though of a different nature, is fully as strong as that of the ground in which sheep have been folded.

tion," whispered he. "Pray pardon me, my excellent friend;" and he pressed Lightfoot's face gently into the bog.

At this moment the deer began to hesitate; to look again around them, and to consult their leaders before they determined upon their course.

The lying concealed in expectation of a doubtful event, and almost within reach of the deer, is one of the most nervous situations imaginable. In running with them there are various things to distract your attention: caution to preserve the wind; prudence to keep your limbs entire in going at the top of your speed down rocky declivities, or amongst large stones concealed in the long ling. Even in creeping for a quiet shot, you are naturally somewhat engaged in ejecting the mud from your mouth, deeming it, perhaps, unpleasant or unwholesome. There is also a sensation when the water enters your shirt breast, which, although not novel, may be termed somewhat interesting. Thus the care bestowed upon your outward man diminishes in some degree the agitation of your mind; but really when you are lying prostrate, in expectation of the deer passing without any effort of your own,—when you hear the trampling, the rush, and the belling, and all this under doubtful auspices, you must be the most odious of all stoics if your pulse beats evenly. We are agitated in such a case —tremendously agitated, we own: our heart trembles within us; our breath comes short; and the whole goddess Diana possesses us. Let those who have cold blood pride themselves on it when they need, and where they need— not now.

See the noble herd are come in view! Na-Shean Tulichean never bore upon his green swells a prouder burthen. The antlers rise and sink over its heights; the hinds and calves pass belling along, whilst we (practising, at least for once in our lives, the virtue of forbearance) feel all the torments that the fabled and thirsty sinner felt as he caught at the flying waters. Yes, the fable may be told of us, and that somewhat to our credit.

And now the great bulk of the herd had passed over the knowes, and were out of sight; still they came on in numbers; but ever as they passed the antlers grew scarcer and scarcer. Tortoise pressed the arm of his companion in silence; at length he removed his hand.

"Now, then, all is safe; follow me."

He sank out of sight over the hill to the west with rapid foot and bent body, and then came in more southwards, within shot of the tail deer, when both sportsmen knelt down on the heather. As the hinds came on, an anxious look was sent to the rear in hopes to descry the points of an approaching antler. At length the horns actually did appear; and Lightfoot, all trembling with eagerness, was clapping his rifle to his shoulder, when Tortoise stayed him, whispering in his ear, "A worthless beastie, my good fellow, let him pass: remember the four-year-old—the enormous monster—the *haud credo:* this is a twin to him." But nothing better came on—nought but rubbish. So not a shot was fired.

They now gave them a little time to get on, and then peeped through the heather-tops at the slope of the green knowes. There they saw the vast herd below them, which had kept increasing their forces as they passed the lower grounds. There might have been some four or five hundred of them altogether.

The deer now began to form into a more compact body. Some looked back, some towards the slaps in the dyke, others to the east and west. Now they drew up on an eminence to the east: they longed for the security of the woods, but were afraid to venture. Sometimes they were about to break to the west, some on the opposite quarter; but at every point they met with opposition. At these critical moments, various were the pushes made by the sportsmen in the rear to each flank of the green knowes in accordance with their motions. Still as they ran they were concealed under the rising ground. Pressed on their flanks, and alarmed on their rear, the woods seemed the only refuge for the herd; and a long string of harts and hinds raced away within shot of some stone dyke that bounded them; the rest of the body lingered behind, as if to ascertain how the experiment would succeed.

Now began the din of arms: two rifle shots echoed through the hollow woods, and two noble harts bit the dust. "That must be the Duke's deed; it is his Grace's usual station; besides it was done so cleverly." Other shots followed, more or less successful, which turned the leaders, and those that came up in the rear sprung high in the air over their fallen comrades, wheeled back, and all again assembled on the flat ground. They now knew that they were beset on all sides, and soon came to a decision. The hinds had hitherto taken the lead; but, pressed as they now were, a more undaunted chief took the command. Stern and determined, a magnificent hart stepped forth from the ranks, and stood singly for a space in all his vast proportion: he towered above the herd, as the Satan of Tasso above the infernal host—

"Si la gran fronte, e le gran corna estolle."

For a few moments he shifted his gaze from man to man; then he made a desperate charge, followed by the rest of the body. It was evident now that they were breaking out in the west; they all swept round behind a low rise of ground, in that quarter,* at the top of their speed.

"Now then, Harry, run low, and do your best."

Down he and Tortoise came upon them, and arrived just in time for the middle of the herd. Two fine harts fell to their rifles. And again, as they raced by the peat-stacks, another party fired upon them; and they came so close to the hill-men that they flung their sticks at them, and had they not given way, would have trampled them to the earth. They now broke back over the moor, and were no longer thought of. It would have required much skill and many hours to get the wind of them again.

"Well, this is a noble day's sport; but you must say nothing about the hind at the castle, Maclaren. To be sure, she will be seen to-morrow at the slaughter-house, and, no doubt, she will have companions of the same gender; but sufficient for the day is the evil thereof; and, indeed, it is of no consequence, for she will make soup fit for the supper of Lucullus,—if you know who he was, Peter?"

---

* This swell of ground is very low, and not far from the wood, and insufficient to mask the deer entirely. I often thought it might be possible to use it to advantage, and now tried it for the first time.

"No, I do not;—was he a Badenoch man?"

"Not exactly; nor had he Badenoch cooks that I ever heard of."

The parties now met, and exchanged greetings and congratulations. There were six first-rate harts slain at the wood, and two lesser harts and two hinds at the peat-stacks. The Duke of Atholl's deer (he had shot three in all) were the largest; for he had ever a quick eye, and an amazing tact in selecting his quarry. One of these was lying on the moor unable to rise, but still alive. It proved to be the large mouse-coloured hart which had escaped the stalkers at Cairn Cherie, and whose fate had been prophesied. A hill-man, unaccustomed to treat with such dangerous animals, went up to him and seized him by the horns without ceremony. An evil deed it was for him; for the stag, tossing up his head, cut him with one of his brow antlers between the eyes, dividing the flesh up his forehead, and giving him a frightful wound. The poor fellow ran up to the Duke, and saying, "Yon was an unco crabbed beast," fell senseless at his feet. He soon recovered himself, however, and was kindly administered unto,—the men deluging his wound with whiskey, which they esteemed a sovereign remedy for all evils under the sun.

Ponies had been kept in readiness to take home the deer; they were a hardy race, redundant in mane and tail, and contemners of the bridle. Amongst these was one known by the name of "Old Blair Pony," who had always the honour of bringing home the Duke's deer. It was an office he delighted in; and he was wont to evince his sense of pleasure by rubbing his muzzle in the blood, and by towzling the beast, as Squire Western has it.

Two or three sportsmen discharged their rifles at the gillies' bonnets, at the distance of a hundred paces, the gillies wisely pulling them off and planting them in the heather, and not standing the shot themselves, as did the Gown-cromb of Badenoch. The light infantry galloped home on their ponies; then followed the shelties, each with a hart corded on his back, with the head and horns uppermost: these were attended by a group of hill-men and gillies, in their kilts and plaided tartans; some urging on the ponies with Gaelic admonitions, others holding the rough lurcher in the leash, and tugging him back rudely as he tried to get a lick of the blood-stained deer. Thus they passed merrily through the storm-beaten forest, winding over the bridges, the dark torrent of the Banavie brawling and toiling below them.

May they enjoy the right good cheer and merry dance that always awaited them at the castle!

Eight harts slain at the wood, and two at Cairn Cherie. By the rood, it was a sufficient work; though the sport had occasionally been much more ample.*

CHAPTER X.

Original Scotch Greyhound.—Fingal and his Retinue.—Bran and Phorp.—Their Death.—The Lurcher —Glengarry's Dogs.—Of Blooding Deer-hounds.—Four-footed Hannibal.—Sir William St. Clair's Dogs.

> " Syr, yf you be on huntynge found,
> I shall you gyve a good greyhounde,
>   That is dunne as a doo;
> For as I am a trewe gentylwoman,
> There never was deer that he at ran,
>   That myght yscape him fro'."
>   Sir Eglamore.—*Metrical Romance.*

The best sort of dog for chasing the deer would unquestionably be the original Scotch or Irish greyhound; but of this noble animal I shall myself say nothing, being enabled, through the kindness of Mr. Macneill of Colonsay, to introduce amongst these pages a dissertation on their race and qualities, put together by him with great research and ability, and accompanied by a recital of a day's deer coursing in the island of Jura. All accounts I have received from Scotland represent these dogs as very scarce at the present day; and I am informed that in Sutherland the last of the race in that particular district was a very powerful animal

---

* It may appear, perhaps, that in the account of this day's drive and the former one, the lion's share of the sport is given to the stalker. It must be remembered, however, that those who go round with the drivers have necessarily the greatest number of chances. Hence Tortoise's success.

belonging to the late Mr. Gordon of Achness. He was killed by a stag about forty years ago, who transfixed him with his antlers against a rock, leaving three deadly wounds on his body.

The traditions of that country have handed down stories to us that prove the great estimation which dogs were held in at very remote periods. One of these traditions, which was current ages before Macpherson's publication, runs as follows:—

Fingal agreed to hunt in the forest of Sledale, in company with the Sutherland chief, his contemporary, for the purpose of trying the comparative merits of their dogs. Fingal brought his celebrated dog *Bran* to Sutherland, in order to compete with an equally famous dog belonging to the Sutherland chief, and the only one in the country supposed to be any match for him. The approaching contest between these fine animals created great interest. White-breasted Bran was superior to the whole of Fingal's other dogs, even to the "surly strength of Luah." But the Sutherland dog, known by the full sounding name of *Phorp*, was incomparably the best and the most powerful dog that ever eyed a deer in his master's forests.

When Fingal arrived in the forest with his retinue and dogs, he was saluted with a welcome that may be translated thus:—

> "With your nine great dogs,
> With your nine smaller, game-starting dogs,
> With your nine spears—
> Unwieldy weapons!
> And with your nine grey sharp-edged swords,
> Famous were you in the foremost fight."

The Sutherland chief also made a conspicuous figure with his followers, and his dogs and weapons for the chase. Of the two rival dogs, *Bran* and *Phorp*, the following descriptions have still survived amongst some of the oldest people in Sutherland. *Bran* is thus represented:—

> The hind leg like a hook or bent bow,
> The breast like that of a garron,*
> The ear like a leaf.

---

* A stout gelding.

Such would Fingal, the chief of heroes, select from amongst the young of his hunting dogs.

Phorp was black in colour, and his points are thus described :—

> "Two yellow feet, such as *Bran* had;
> Two black eyes,
> And a white breast;
> A back narrow and fair,
> As required for hunting;
> And two erect ears of a dark brown red."

Towards the close of the day, after some severe runs, which, however, still left the comparative merits of the two dogs a subject of hot dispute, *Bran* and *Phorp* were brought front to front to prove their courage; and they were no sooner untied, than they sprang at each other, and fought desperately. *Phorp* seemed about to overcome *Bran*, when his master, the Sutherland chief, unwilling that either of them should be killed, called out, "Let each of us take away his dog." Fingal objected to this; whereupon the Sutherland chief said, with a taunt, that "it was now evident that the Fingalians did not possess a dog that could match with *Phorp.*"

Angered and mortified, Fingal immediately extended "his venomous paw," as it is called (for the tradition represents him as possessing supernatural power), and with one hand he seized Phorp by the neck, and with the other, which was a charmed and destructive one, he tore out the brave animal's heart.

This adventure occurred at a place near the march, between the parishes of Clyne and Kildonan, still called Leck-na-con, the Stone of the Dogs, there having been placed a large stone on the spot where they fought. The ground over which Fingal and the Sutherland chief hunted that day is called Dirrie-leck-con. Bran suffered so severely in the fight, that he died in Glen Loth before leaving the forest, and was buried there. A huge cairn was heaped over him, which still remains, and is known by the name of Cairn-Bran.*

---

* Mr. Grant of Corrymony, in his work on the Gael, relates a tradition somewhat similar to the above, and which may have been drawn from the same sources; but it differs from it in stating that Bran was the victor, and in the omission of his death.

Not being in possession of any of the celebrated race of the original Scotch greyhound, which are now, indeed, very rare, and finding that all the dogs in the forest of Atholl were miserably degenerate, I bred some litters from a foxhound and a greyhound, the **foxhound** being the father. This **cross** answered perfectly: indeed, I was previously advised that it **would do so** by Mr. John Crerer, who, after having tried various crosses for **sixty years**, found this incomparably the best. Neither **of these animals** themselves would have answered; for the **greyhound cannot stand** the weather, and **wants courage to that degree, that** most of them will turn **from a fox when they come up to him, and** see his grin, and **feel his sharp teeth**; nay, they will scarcely go through a hedge **in** pursuit of a hare **till after** some practice. Besides, they have **no** nose, and run entirely by sight; so that when the **hart** dashes into a deep moss or ravine, the chase **is** over, and the dog stops, and stares about him like a born **idiot as** he **is**.

The foxhound **is** equally **objectionable**; he has not sufficient **speed, gives** tongue, **and hunts too much** by scent: in this **way he** spreads alarm **through the** forest; and if **you turn him loose, he will amuse himself all day** long, and **you will probably see him no more till he comes home at night to his kennel**.

All **these objections are** obviated by **the** cross **between the two. You get the** speed of the **greyhound, with just enough of the nose of the foxhound to answer your purpose. Courage you have in** perfection, for most dogs so bred **will face anything**; neither craggy precipices, **nor rapid streams, will** check their course; they run mute, and **when they are put** upon the scent of the hart, they will **follow it till they come up to** him; and, again, when he **is out of view**, they **will** carry on the scent, recover him, **and** beat the best greyhound to fits: **I** mean, of course, on forest ground.

The present Marquis of Breadalbane had two dogs of **this** description, **Percy** and Douglas, which were bred by me. As they were my very best upon scent, I gave the late Duke of Atholl the use of them every season, to bring cold harts * to bay, in which they were wonderfully success-

---

* A *cold hart* means one that has not been wounded.

ful; for if they were fairly laid on, no hart could escape them. They are now nine or ten years old; and his lordship informs me they are still able to bring the stoutest hart in his forest to bay, and are altogether perfect.

These dogs, in point of shape, resemble the greyhound; but they are larger in the bone, and shorter in the leg: some of them, when in slow action, carry their tails over their backs, like the pure foxhound. Their dash in making a cast is most beautiful; and they stand all sorts of rough weather.

As the above is, I think, the best cross that can possibly be obtained for the modern method of deer-stalking, so it should be strictly adhered to: I mean that, when you wish to add to your kennel, you must take the cross in its originality, and not continue to breed from the produce first obtained; for if you do this, you will soon see such alarming monsters staring around you, as the warlike Daunia never nourished in her woods and thickets, or as cannot even be surpassed by the sculptured ones at the villa of Prince Palagonia, near the shores of Palermo.

The late celebrated sportsman, Glengarry, crossed occasionally with a bloodhound instead of a foxhound: his famous dog Hector was probably bred in this way; and I believe Maida, the dog he presented to Sir Walter Scott, had also a distant cross of the bloodhound in him. Two of these small bloodhounds he generously gave to me, though he was chary of the breed; but they ran away from my kennel, and were unfortunately lost.

A cross with the bull-dog was once tried in the forest of Atholl, to give courage; but the produce was slow, as might have been expected; and the thing was overdone, for they all got killed by attacking the deer in front. High-couraged dogs, indeed, of every breed, are subject to accidents: they get wounded, and even killed, by the harts; are maimed for life, or meet their death by falling over precipices in their reckless pursuit, particularly in rounding a corner.

It is very seldom that the deer themselves suffer from precipitous falls, being well acquainted with their ground, and studious in selecting it. Once, however, when I was out, it happened that a hart, being wounded by me, and chased by one of my hounds, came to a very high and steep declivity by the river Mark, not far from Glen Tilt. Being pressed closely by the dog, he went down it upon his hind quarters, preserving his position in the rush in a most wonderful manner, at a time when I expected he must have fallen headlong, and met with inevitable and instant death. The dog just saved himself in the scramble, and had barely power to draw back, pausing for a moment at the edge of the precipice, with his fore legs extended, and horror in his looks. The hart was not dead, though terribly mangled. I got to him with difficulty, by going some little distance round, and swinging down from rock to rock by means of the impending birches.

For my own sport I seldom turned my dogs loose after cold harts, only doing so when I was endeavouring to bring such to bay for the Duke of Atholl's sport. Thus being put upon the scent of wounded deer only, they stuck to the blood.

The hill-man or gillie who leads the dogs should be a very steady clever fellow, and, moreover, a strong man; for the dogs are so eager and powerful, that he who has them in the leash is frequently pulled head over heels, when he runs down hill with them. All their tackle should be strong, and regularly inspected every morning, lest the dogs should break loose, disturb the cast, and ruin your sport for the day. Guard against all carelessness of this sort.

The dogs should be led about a hundred yards behind the deer-stalker; and the leash-man should stop when he stops, and stalk him as he stalks the deer. Should the herd come in sight, he had better get them to lie down in a hole if possible, and put his handkerchief over their eyes, or they will be apt to struggle or whine, and do irreparable mischief. After the shots are fired, it is the man's duty to run up with them in the leash, some few degrees quicker than the American vessel, which was unsuccessfully chased by a flash of lightning. He then gives them up to the forester, who lays one of them on, if there is occasion; one good dog being quite sufficient to bring a wounded hart to bay.

It may sometimes be requisite to slip a dog immediately:

for instance, if a hart is shot through the loins he will fall prostrate, spring up again suddenly, and baffle a good dog afterwards. There are certain other cases also when despatch is necessary; but, generally speaking, it will be prudent to take time; and the party had much better lie down in the heather, and keep an eye on the wounded deer through the telescope. If he is slightly wounded, it is of no use to send a dog after him at all, unless he is alone; for he will get into the middle of the herd, and keep there with enduring pertinacity; and the thing will just end by your losing him, and bringing a singularly lean hind to bay; throwing away, by a moderate computation, two or three precious hours, and with them, perhaps, your remaining chance of sport for the day: but, on the contrary, if he is badly wounded, and you do not press him on, he will gradually get worse and worse, and fall out from the parcel, when you will have him safe enough. The forester should then pass the track or taint of the herd, and either lay the dog on the scent, or put him in sight of the quarry, and he will soon bring it to bay, if he is worthy of his ancestors. But I have touched upon this subject before.

Some sportsmen are accustomed to give their dogs portions of the deer's liver when he is gralloched; but, after having blooded them once or twice, to enter them, I do not think the custom should be continued, a dog's love for sport being independent of eating; for pointers will hunt gallantly all day long, and they are never permitted to touch their game, nor even to run after it. Harriers, likewise, will persevere from morning till night, and yet the hare is always preserved for the table, if possible,—more particularly in a subscription pack.

My objection to the system lies principally in the two following reasons: the first is, that a dog can never run a second chase properly after having been so fed; the second, that when he has a deer in a wounded and dying state, he is apt to help himself from the haunches before you have time to come up. A lurcher once damaged my sport in this villainous manner. I had wounded a deer which came out unexpectedly from Glen Croinie, against my wind, during a heavy mist. A dog was slipped and laid on the scent. For a long time, we could neither hear nor discover the bay: at length we came suddenly upon it, if bay it might be called. The dog had taken steaks from the living haunches, after the fashion of Abyssinia, and was already amazingly turgid. His name was Hannibal.

> "Expende Annibalem, quot libras in duce summo Invenies."——

I gave him a pretty considerable drubbing for this his luxurious propensity; but even under the lash, it was sometime ere

> "La bocca sollevò dal fiero pasto
> Quel peccator."——

After this perpetration, I changed his name, by a very easy transition, from Hannibal to Cannibal; but Hannibal or Cannibal, I never suffered him to pass the Scotch alps with me a second time.

There is an interesting story mentioned in the notes of the "Lay of the Last Minstrel," taken from a manuscript "History of the Family of St. Clair," which is so apposite to this subject, that I cannot forbear transcribing it.

It seems to prove that the chief reliance for sport was formerly placed in the dogs, who were accustomed to pull down and kill deer without any aid from the huntsmen; and that nobles, and even kings, prided themselves upon the fleetness and courage of their hounds.

"King Robert Bruce," says Augustin Hay (canon of St. Généviève), "in following the chase upon the Pentland Hills, had often started 'a white faunch deer,' which had always escaped from his hounds; and he asked his nobles, who were assembled around him, whether any of them had dogs which they thought might be more successful. No courtier would affirm that his hounds were fleeter than those of the king, until Sir William St. Clair of Roslin unceremoniously said, that he would wager his head that his two favourite dogs 'Help' and 'Hold' would kill the deer before she could cross the march-burn. The king instantly caught at the unwary offer, and betted the Forest of Pentland-Moor against the life of Sir William St. Clair. All the hounds were tied up, except a few ratches, or slow hounds, to put up the deer; while Sir William St. Clair, posting himself in the best situation for slipping his dogs, prayed devoutly to Christ, the blessed Virgin, and St. Katherine. The deer was shortly after roused, and the dogs slipped, Sir William following on a gallant steed to cheer them. The hind, however, reached the middle of the brook, upon which the hunter threw himself from his horse in despair. At this critical moment, however, Hold stopped her in the brook; and Help coming up, turned her back and killed her on Sir William's side. The king, descending from the hill, embraced Sir William, and bestowed on him the lands of Kirkton, Loganhouse, Earncraig, etc., in free forestrie."

The tomb of this Sir William St. Clair, on which he appears sculptured in armour, with a greyhound at his feet, is still to be seen in Roslin chapel.

CHAPTER XI.

Occupation of Forest Lodge.—Autumnal blasts.—Sullen fuel.—The sport begins.—Deer stalker distressed.—A sharp walk.—Lying in ambush.—The fatal spot reached.—Herd in jeopardy.—Peter Fraser's humanity.—His penmanship.—The lament.—The moors.

> " Farewell to the Highlands, farewell to the north,
> The birthplace of valour, the country of worth:
> Wherever I wander, wherever I rove,
> The hills of the Highlands for ever I love."
>
> A. MACDONALD.

I HAVE elsewhere observed, that I have forborne to recount my most successful days on the hills, as not always being fraught with any very marked interest; I now, however, proceed to relate the events of one auspicious day, which, as it was my last, so it was, perhaps, my best. It will prove that the method of stalking deer in quick time, where the forest is sufficiently extensive to admit of such sport, is frequently accompanied by the most abundant results. Three or four skilful attendants you must have for this purpose: one to assist you in stalking and carrying the spare rifles; two more to coax the deer towards you, that is, one on either flank, at vast distances; and another to hold the dogs. This is generally a sufficient force for the Forest of Atholl; but a fifth man would be useful in a west wind, to leave at the mouths of the glens, and keep the deer from going north, which they are apt to do at such a time.

The Duke of Atholl and the shooting parties had all left Blair, and the occupation of the forest was indulgently given to the writer of these pages, accompanied with the most pleasant of all commissions; namely, that of an injunction to kill as many harts as possible, and to take possession of Forest Lodge, the best situation for sport in the whole domain. Captain Stewart of Murthly, an active and skilful deer-stalker, had permission to take two or three days' diversion from his quarters at Blair; but, as better sport might be expected in Glen Tilt, Tortoise took the liberty of asking him to repair to Forest Lodge, and he came accordingly. It was arranged over-night, that the Captain should take the cast east of Glen Croinie, which had not been disturbed for a long time, and that his friend should occupy the ground on the west of it.

The season for deer-shooting was now nearly terminated, and the brightness of the autumnal days had suffered some diminution. The sun withdraws its beams from the secluded Glen of the Tilt early in the evening, and returns only in partial gleams, till the day is spread in full splendour on the summits. Thus the air there is dank and chill; the leaves soon fall from the old weather-beaten birches, and here and there they already stood amongst the grey rocks, in all the nakedness of winter. Some, in more sheltered spots, perhaps, retained their leaves longer, half withered as they were, and shivering in the bitterness of the blast. The weather was soon expected to break up, and the silence of the great waste to be followed by the howling of the storm and the roaring of the cataract.

It was in this dubious season that our sportsmen were stationed in Glen Tilt: as they pass the night in the comfortable tent beds, the peats just expiring on the hearth-stone, they hear, amid broken slumbers, the wind rushing along in fitful gusts, and the rain drops striking fiercely against the panes of the casement; shortly, perhaps, they cease; the moon flashes out for awhile, and her light strikes clear against the furniture of the little chamber; then the black clouds hurry along, blot out her orb, and leave the cottage and glen in darkness and in mystery: thus the night wanes; and amid these contentions of nature, the blustering waters of the Tilt sound loud and continuous: their voice may be somewhat smothered for the moment by the noise of the wind; but, in proportion as that abates, the eternal roar of the torrent swells forth again in all its turbulence.

At length the morning began to dawn, and Tortoise went forth, and paced about the Lodge, that he might endeavour to satisfy himself as to the weather. The wind was still fair; but the air was raw and wintry, and a thick vapour rested upon the mountain ranges. Well, that might pass away:—and now for the morning meal. Who can tell how often the bellows was applied to the sullen fuel, and how reluctant the peats were to confess the slightest capacity for a flare-up? At length, after much coaxing and perseverance, behold a faint ignition; thus things began to brighten, and breakfast was soon put upon the table, warm and redundant. But the less we say about the viands the better; we are rather shy of mentioning such things in detail. We should despair, indeed, of making ourselves understood as to the extent to which the principle of eating may be carried on by the minions of the mountain.

We may be allowed to hint our opinion, however, that those were rational times, when maids of honour drank ale and ate chines of beef at sunrise, with true feminine alacrity. Well, let this pass. Our temptations were vehement, we own, but we do not say we indulged them; and, having before discussed this subject, it does not become us to resume it. We are already on our shelties, replete or empty, it matters not.

And now the ponies plant their feet heavily, and go winding and tugging up the mountain. Captain Stewart strikes off with his men to the right. "Good sport to you, Captain, and a steady hand!"

Tortoise aspires at once to reach the nearest sky-line; the

bogs soon become deep, and the pony is sent back to the regions below. Onward he strides on foot, lessening to the sight by degrees, till he is dimly seen from the glen, and soon entirely lost in the mountain mist. As no operations can be carried on during such an impediment to the view, the party sit down in a little hollow near the summits, where a small burn creeps lazily through the mosses. But the vapours rise speedily, and form into small clouds, that begin to dapple the distant mountain-peaks: onward move the party cheerily; the day promises fairly; the wind is propitious: Care sails scowling with her hollow eyes through the vapour, and leaves our riflemen with the comfortable prospect of a fair field for operations.

The sport began unexpectedly; for a few deer, that could not be seen during the mist, broke out suddenly from a hollow towards the east, at the back of the Grianan-Moir, and raced away towards Cairn-chlamain. They were at an awful distance; but as the course of the leading ones was decided, and the tail ones in the hollow were out of sight of the rifleman, he made a dash forward, and thus gained considerably upon the spot of their crossing; so that when these latter began to appear, he took a long shot at a hart, which was evidently struck by the ball.

"Never heed him, Peter; forward, forward, man."

"Why, sure, then, we mun stop and tak' tent o' the deer?"

"No, no; no such thing. Here, Maclaren, take Percy; run forward, and hold the deer at bay. Come along, Peter, more deer will join them, and we shall have them again as they come out of the mouth of Glen Croinic."

Away they dashed at the top of their speed, at least Tortoise most assuredly did so; but as he made "gallant show and promise of his mettle, so, like a deceitful jade, he sank in the trial." What, dead beat! He whom Maga in former times, and in her flattering mood, extolled for feats on river, mountain, lake, and moor; he dead beat? Alas! yes, most certainly, most undeniably so, and blowing like a grampus. The way was short: but what will not pace effect? Some how or another, however, he held on without being much the worse for it.

Thus he contrived to reach the mouth of the glen in piteous plight, and something in the attitude of the Austrian spread eagle, just as the tail deer were sinking the hill down to the culreach. A shot was fired, and it was fortunately a clean one: a fine stag fell dead on the spot.

"Now halt, my good fellows, and let us watch the deer."

They saw them pass over the hill to the west, and lost them for some time in the glen below. At length they crossed the river Mark and re-appeared, ascending the opposite mountain just south of Cairn-cherie; slowly did they climb the brae, and, being completely tired, lay down on the moss some way up the hill.

"Very well, gentlemen, we will talk to you by and bye."

"Now, Fraser, whilst Sandy is gralloching this deer, do you go and seek the bay of the other."

Nor was this trouble a toilsome one, for Percy held at him in the moss under the grey stones of Cairn-Chlamain; and a ball was soon sent through his head.

"Now, then, take up the rifles, lose no time, and follow me, Peter."

"Why, what can we do? thae beasties are in sight o' a' the glen, and we canna pass the Mark burn at ony gate."

"It will be a long round, and a toilsome one; but you did not get your bonny wife, you know, Peter, by means of a faint heart. Here, Maclaren, do you remain on this brae (they had advanced some way), and when with your glass you see us fairly above the deer, wait for our signal; we will draw breath a space before we give it. But when you do see it, put the deer over to us in your very best style. Now, Fraser, hard work as it is, this is our only chance; but you are never tired, blown or daunted; it is no use to go back towards the east, the ground is all disturbed there; so we must take a long round by Coir-na-minghie, and cross the low ground out of sight, where we can go up Cairn-cherie, and get above them, and then let them look to themselves."

All this was done at their best pace: after a long, I will not say a toilsome, circuit—the excitement they felt rendering them insensible to fatigue,—a close approximation to the fatal spot was gained. They had the deer below them, that was certain; but it was necessary to ascertain their precise situation before they were started, and **not to lose** sight of the points of their horns whilst they were running, otherwise a complete failure might be anticipated. For in such a case they might come out behind the sportsman whilst he was running forward, get his wind, bolt out of the cast, and thus be lost to him for the day; or they might cross the ground out of distance, or go straight forward out of sight. Success, in short, in such case, would depend upon mere accident; so the proper tact was observed; they kept well behind them, and peeped and crawled for some time, till they discovered a hind. She was lying down in the moss, shaking her head and flapping her ears, as if to keep off the flies. Every now and then she looked up and gazed about her with expanded nostrils, as if to search for some taint in the air. She was evidently the leader of the parcel, and the harts were sure to wait upon her movements.

Tortoise, Peter Fraser, and Thomas Jamieson now crept back, and went on a little till they got to some ground, under cover of which they were able to proceed in a more comfortable attitude. They then got on cautiously to the south-west, and after some curious windings, and certain dabblings in bogs and water-courses, they laid themselves down prostrate in the heather, through bunches of which they had a glimpse of the cautious sentinel. Jamieson, who prudently lagged behind, was then motioned to give the signal, which was the exhibition of his shirt by the unbuttoning of his waistcoat—an object discoverable by the glass at a very considerable distance.

No sooner had the signal been observed by Maclaren, who it will be recollected was on the opposite mountain, than he rose up and came forward in the direction of the herd; as he advanced slowly, the hind stood up, and the horns of the stags below her began to appear to the sportsmen one after the other, and presented a most tantalising spectacle. These fine fellows were at a very considerable distance, but the rifleman completely commanded their position.

After a little shifting and advancing on the part of Maclaren, and continued gazing and observation on the side of the deer, the latter began to draw forward a little, but soon halted, as if to ascertain whether a retreat was absolutely necessary; having at length judged it to be so, they moved on leisurely with a few hinds in front to a notch in the hill, where the ascent was the least fatiguing to them; the hinds sank into this hollow, went forward up it, and were lost sight of in a few moments. The rest of the herd followed them; the sportsmen then rose up warily, and got forward also by a semicircular movement, running under cover of rocks and moss-hags, with sufficient rapidity to bring them within distance as the deer crossed in front of them.

They arrived just in the nick of time, and found themselves about a hundred yards from the herd as it swept by. The bodies of the harts were a fair and inviting mark, though their legs were hid,—the selection was promptly made, and two first-rate stags fell dead upon the spot; the third ball also had evidently hit the mark. Away ran Peter Fraser, whilst Jamieson loaded the rifles, and just glancing at the two victims as he passed them, peeped over the next ridge of the hill, when he suddenly tossed his arms aloft, like Gilpin Horner, and pranced forward to a third deer which lay dead beyond him.

It takes up a considerable time to clean three deer and prepare them properly, so that during this operation the herd had leisure and opportunity to get forward and select their own ground, which they did, by going into Glen Dirie, and moving along the steep stony tracks on the western face of Ben-y-venie.

"Here, Sandy, another glass of Loch Rannoch, the Doch-an-dorroch, ye ken; off with it. So now go up Ben-y-chait, taking care to cross the glen out of sight of the deer, and to keep them from the west. We will go forward right up Ben-y-venie."

Sandy Macintosh was a capital fellow of the antelope sort, and put out his long legs nimbly, so that he was quickly on his ground, as also was the rifleman. The deer were soon discovered winding among the crags below; and keen Sandy was so alert and judicious in his motions, that he kept them on that precipitous side of the mountain.

"By heavens, he has turned them up our hill again, and they are coming up the steeps at the old place! Forward, forward, run low, low; we shall have them again to a certainty."

He did indeed have them with a witness, and came right up with a string of them, running immediately below him at an easy distance. Go which way they chose they could not now escape him; a vast hollow of the hill side lay beneath, fully exposed to his view; so he stood on the commanding crags, without the slightest attempt at concealment, and fired two shots in rapid succession. One hart fell dead on the spot, and another went away wounded.

"Murder, murder! O Lord, murder! Haud yer han'; haud yer han'; we canna tak' tent o' a' thae deer.".

And Peter Fraser held the third rifle with a firm grip, and refused to give it up. But a sharp tug or two, and a sudden and unexpected twist from Tortoise, soon released it from his grasp.*

"Aweel, aweel; haud to yon muckle deer then, awa to the wast. There, there *(pointing)*."

Down he dropped instantly to the rifle; and away went Tortoise after the wounded stag. A dog was properly slipped, who ran a beautiful chase all down the steeps of Ben-y-venie towards the river Mark. There the helpless animal stood at bay, and received his death-shot. He fell in a secluded spot, below some rocks and birch trees, where he was gralloched and washed out; his head was turned back on his shoulder, according to custom, and peats were put upon it to keep his eyes from the great bird; nor did they neglect to tie the black flag on his horns, which, waving in the night air, might scare away the raven, and baulk him of his prey.

The herd passed forward, and Tortoise held his glass to them, but discontinued the pursuit, although they were still before him in his cast.

---

* The immediate attendant on the deerstalker holds the spare rifles, and gives them one after the other to the sportsman, as he fires them in succession. The gunstocks got much battered in Tortoise's service, as he generally flung down each rifle as soon as he had discharged it—rock or moss, it took its chance.

The events of this day may be summed up in the words of Peter Fraser, which I extract from a letter written by him, now lying before me, and which he sent to Dunkeld, for the purpose of communicating this remarkable day's sport.

"The deer went on to Beinn-a-Weadhounedh,* and before we was done with the aforesaid hill, Mr. S—— had his day's sport finished—eight fine harts. This was done early in the afternoon; and he wished to carry on further, but I got him advised to go home to Forest Lodge." †

This was my last day in the forest of Atholl. The scene, alas, soon changed, and mourning followed on its rear. In the midst of joy comes sorrow—the dark, the inevitable cloud, which had been almost imperceptibly gathering, at length burst over us. The solemn bell of the old Cathedral struck duly, and the sound bore the lament through the hollow woods and glens, and fell heavily upon our hearts; the waters rolled on, and the pines waved their green heads, but all was void and desolate. That intellectual light, which shone on the vast domain,—which, acting on a Roman scale, gave employment and a maintenance to thousands,—which spanned the broad waters of the Tay with a magnificent bridge, and spread immense forests over wastes heretofore unproductive—which was evermore successfully exerted for the happiness of family, friends, and dependants, and the prosperity of the country at large—that light—that master-mind, was suddenly withdrawn from us, and the kindest heart that ever warmed human bosom ceased to beat. Sorrow sat brooding in the halls of the great; and the rough Highlander, as he walked silently in the gloom of

---

* The Gaelic appellation for Ben-y-venie.

† The Duke of Atholl was so liberal in his presents of venison, and his hospitality so great, that no supply I was ever able to afford him could exceed his desires, so that he rejoiced in a day of this description, and would listen to the details with great interest. Some apology would otherwise be necessary for my slaughter on this and other days nearly similar to it. The chief point consists in selecting the best harts, and passing by the inferior ones. This was held to be the test of a good sportsman. In grouse-shooting, except I was enjoined to do otherwise, I always limited my sport to twenty brace a day, though in a good season I think I could have killed four or five times that number; but I never had any pleasure in destroying game for which there was no immediate demand. Peter Fraser still acts as deer driver in the forest of Atholl.

the glen, paused, and drew his sleeve across his eyes, as he thought on his departed chief.

The bitterness of that hour is now past, and a new dawn breaks over the mountains. The gallant young heir returns to his native hills and floods, radiant with youth and promise; his people accept the omen.

Proceed noble chieftain, and fulfil your great duties like him who is gathered to the tomb of his fathers; and may his mantle sit gracefully on you. May happiness and the well-earned love of your dependants wait upon your footsteps; thus the glory shall shine on your brows, and depart not from the halls of your ancestors.

> ——— " Si quâ Fata aspera rumpas,
> Tu Marcellus eris ———."

THE MOORS.

By the Hon. T. H. Liddell.

The moors, the moors, the bonny brown moors,
Shining and fresh with April showers!
    When the wild birds sing
      The return of spring,
      And the gorse and the broom
      Shed the rich perfume
      Of their golden bloom,
'Tis a joy to revisit the bonny brown moors.
Aloft in the air floats the white sea-mew,
And pipes his shrill whistle the grey curlew;
And the peewit gambols around her nest,
And the heath-cock crows on the mountain's crest;
And freely gushes the dark brown rill,
In cadence sweet from the lonely hill;
Where, mingling her song with the torrent's din,
As it bubbles and foams in the rocky linn,
Twitters and plunges the water-crow
In the pool where the trout are springing below;
And the lambs in the sun-shine leap and play
By their bleating dams in the grassy brae,
With a withered thorn for their trysting place,
To mark the goal where their foot-prints trace
The narrow course of their sportive race.
Oh! know ye the region in spring more fair
Than the banks and the glens of the moorland bare?
The moors! the moors! the fragrant moors!
When the heather breaks forth into purple flowers!

> When the blazing Sun
> Through the Crab hath run,
> And the Lion's wrath
> Inflames his path,
> What garden can vie with the glowing moors!
> The light clouds seem in mid air to rest
> On the dappled mountain's misty breast,
> And living things bask in the noon-tide ray,
> That lights up the summer's glorious day;
> Nor a sough of wind, nor a sound is heard,
> Save the faint shrill chirp of some lonely bird—
> Save the raven's croak, or the buzzard's cry,
> Or the wild bee's choral minstrelsy,
> Or the tinkling bell of the drowsy flock,
> Where they lie in the shade of the caverned rock:
> But when the last hues of declining day
> Are melted and lost in the twilight grey,
> And the stars peep forth, and the full-orbed moon
> Serenely looks down from her highest noon,
> And the rippling water reflects her light
> Where the birch and the pine-tree deepen the night:
> Oh! who but must own his proud spirit subdued
> By the calm of the desert solitude:
> So balmy, so silent, so solemnly fair,
> As if some blest spirit were riding the air,
> And might commune with man on the moorland bare!

> The moors! the moors! the joyous moors!
> When Autumn displays her golden stores;
>   When the morning's breath
>   Blows across the heath,
>   And the fern waves wide
>   On the mountain's side,
>   'Tis gladness to ride
> At the peep of dawn o'er the dewy moors!
> For the sportsmen have mounted the topmost crags,
> And the fleet dogs bound o'er the mossy hags,
> And the mist clears off, as the lagging sun
> With his first ray gleams on the glancing gun,
> And the startled grouse, and the black cock spring
> At the well-known report on whirring wing.
> Or wander we north, where the dun deer go
> Unrestrained o'er the summits of huge Ben-y-gloe;
> And Glen Tilt, and Glen Bruar re-echo the sound
> Of the hart held to bay by the deep-mouthed blood-hound,
> And the eagle stoops down from Schechallien to claim,
> With the fox and the raven, his share of the game.
> But a cloud hath o'ershadowed the forest and waste,
> And the Angel of Death on the whirlwind hath passed

And the coronach rings on the mountains of Blair,
For the Lord of the woods and the moorlands bare.

The moors! the moors! the desolate moors!
When the mist thickens round, and the tempest roars!
    When the monarch of storm
    Rears his giant form
    On some rock-built throne
    That he claims for his own,
To survey the wild war on the desolate moors!
For the winds are let loose, and the sound is gone forth
To awaken the troops of the frozen north!
And the lightning, and hailstone, and hurricane fly,
At a wave of his arm, through the dark rolling sky;
And his footsteps are trampling the fog and the cloud,
That envelop the earth in a funeral shroud;
And the sheep and the shepherd lie buried below
The wide-spreading folds of his mantle of snow;
And the breath of his nostrils encumbers the wood;
And his fetters of crystal arrest the flood;
And he binds in its fall the cataract,
And makes level the gulfs of the mountain tract;
Till his work is complete,—and a dread repose
Broods over a boundless waste of snows;
And the wild winds bewail in whispers drear
The decay and death of the by-gone year.

CHAPTER XII.

A DESCRIPTION OF THE HIGHLAND DEERHOUND, WITH AN ACCOUNT OF A DAY'S DEER-COURSING IN THE ISLAND OF JURA.

[Communicated by ARCHIBALD MACNEILL, *Esq., of Colonsay.]*

Dogs of Ancient Britain.—Irish Dogs sent to Rome.—Early Scottish Dogs.—Sculptured Stones at Meigle.—The Miol-chu.—The Mastiff and Greyhound.—Recreation of Queen Elizabeth.—Dogs of Epirus.—Irish Wolf-dog.—Proportions of a Deer-hound. —Failure of Crosses in Breeding.—Deer Dogs of Colonsay, and Dimensions of Buskar.—Expedition from Colonsay.—Cavern Scene.—Wild Scenery in Jura.—Stag Discovered.—Stalking Him.—The Start and Course.—His Death.—Speed and bottom of Deer-hounds.—Decay of the Ancient Race.

"Canis venaticus, celerrimus, audacissimusque non solum in feras sed in hostes etiam latronesque præsertim, si dominum ductoremve injuria offici cernat, aut in eos concitetur."—BOECE.

IT is not a little remarkable that the species of dog, which has been longest in use in this country for the purposes of the chase, should be that which is least known to the present generation of naturalists and sportsmen. While we are presented with delineations and descriptions of every race of dog, from the mastiff down to the pug, we find no writer of the present day who speaks with any degree of certainty as to the size, colour, or appearance of the deerhound, once so highly prized, and for a great period of the history of this country, the only dog fitted for the sports of the field. One would naturally have thought that the gigantic, picturesque, and graceful form of this animal (the constant attendant of nobility), would have insured for the present generation a faithful description of its appearance and habits, but it is to be feared that none such has been transmitted to us, and that to the effusions of the bards, and traditionary tales of former days, we are chiefly indebted for any idea of the perfection to which this breed at one time attained in this country.

From modern writers we learn nothing further than that such a race of dogs at one time existed in Ireland, that they were of a gigantic size, and that they are now extinct.

One great obstacle in the way of investigating the history of this dog has arisen from the different appellations given to it, according to the fancy of the natives in different parts of the country, of Irish wolfdog, Irish greyhound, Highland deerhound, and Scotch greyhound.

But for these apparently distinctive designations, sufficient information would probably have been recorded regarding a breed of dogs really the same, and in such general use throughout the different parts of the kingdom.

That dogs resembling the greyhounds of the present day were known in this country as early as the third century we have ample proof from the writings of Roman authors, and, in particular, from the works of Nemesianus and Gratius. In his Cynegeticon Gratius mentions two distinct breeds of dogs as natives of England, the one termed Molossus, which is supposed to have been the mastiff, and the other Vertraha, which, from the description, seems to correspond, in many points, with the greyhounds at present in use in this country.

Nemesianus gives the following description of these dogs:—

> "Sit cruribus altis,
> Sit rigidis, multamque gerat sub pectore lato
> Costarum sub fine decenter prona carinam
> Quæ sensim rursus siccâ se colligat alvo,
> Renibus ampla satis validis, deductaque coxas,
> Cuique nimis molles fluitent in cursibus aures."

And again he says,—

> "Divisa Britannia mittit
> Veloces, nostrique orbis venatibus aptos."

From the same authorities we learn that the mastiffs of England were highly prized by the Roman emperors, and were used by them for the combats of the amphitheatre.

It also appears from Symmachus, that in the ourth century a number of dogs of a great size were sent n iron cages from Ireland to Rome, which were probably used for the same purposes; and as the mastiff was purely an English dog, it is not improbable that the dogs so sent were greyhounds, particularly as we learn, from the authority of Evelyn and others, that the Irish wolfdog was used for the fights of the bear-garden.

How and when this species of dog came to be denominated greyhound is a point on which naturalists are not agreed. Some derive the appellation *grey* from Graecus, whilst others, as Jn. Caius, derive it from *gret*, or great. Without pretending to determine this point, it may be suggested, as not improbable, that the name is derived from the colour (which is still the prevailing one of these dogs in the remote districts of Scotland), particularly as we find them described as *Cu lia*, or grey dog.

Whatever may have been the origin of the name, there is little doubt as to the antiquity of a species of dog in this country bearing a great resemblance in many points to the greyhound of the present day, and passing under that name, though evidently a larger, nobler, and more courageous animal.

Among the oldest Scotch authorities are some sculptured stones in the churchyard of Meigle, a village of Perthshire. These stones represent in relief the figures of several dogs, which bear so strong a resemblance to the Highland deerhound as to leave no doubt that they are intended to represent this species. The date of this sculpture is considered by antiquaries, and in particular by Chalmers, to have been previous to the introduction of Christianity, and as early at least as the ninth century.

These, though probably the earliest, are by no means the only stones on which representations are given of these dogs. On many others of great antiquity to be met with in different parts of the country hunting scenes are represented, in which the same species of dogs are introduced in full pursuit of deer.

Among the Anglo-Saxons, with whom the wild boar, the wolf, and the hart were constant objects of sport, no dogs were so highly prized as the original race of greyhounds.

When a nobleman travelled, he never went without these dogs. The hawk he bore on his wrist, and the greyhounds who ran before him, were certain testimonials of his rank; and in the ancient pipe-rolls, payments appear to have been often made in these valuable animals.

In the 11th. century, so greatly were greyhounds in estimation, that by the forest laws of Canute the Great, no person under the rank of a gentleman was allowed to keep one.

At this period, and until after the Norman conquest, the chase was always pursued on foot; the Normans having been the first to introduce the mode of following their game on horseback.

It is obvious, from the rough and uncultivated state of the country, and the nature of the game which was then the object of the chase (viz., deer of all sorts, wolves, and foxes), that the dogs then used would be of a larger, fiercer, and more shaggy description than the greyhounds of the present day, which are bred solely for speed, and have, by modern culture and experimental crosses, been rendered, in all probability, a swifter animal, and better suited for coursing the hare in a level country.

As cultivation increased, the game for which the deerhound was particularly suited gradually diminished, and the improvement in agriculture in England being more rapid than in the sister kingdoms, the diminution of deer and wolves was proportionally great. The deerhound, consequently, in that country, degenerated from want of attention to its peculiar characteristics, and gradually merged into the greyhound of the present day.

In Scotland, Ireland, and Wales, red deer continued to be the objects of the chase till a much later period than in England; and as from the rugged and uncultivated state of these countries the game could only be followed on foot, it was necessary to use that species of dog which would enable the sportsman to view and enjoy the chase.

At an early period, the name by which these dogs were known in these countries was the same, viz., the Celtic one of *Miol chù*, which signifies a dog for the pursuit of wild animals, though this term is now applied generally to all dogs of the greyhound species.* The following description of the miol-chù has been handed down for generations, and is quite as minute, and at least as old, as the well known one of the book of St. Alban's:—

> "Sud mar thaghadh Fionn a chù
> Suil mar airneag, cluas mar dhuileig,
> Uchd mar ghearran, speir mar choran,
> Meadh' leathan, an cliabh leabhar,
> 'San t-alt cuil fad bho'n cheann;"

which may be translated thus:—

> An eye of sloe, with ear not low,
> With horse's breast, with depth of chest,
> With breadth of loin, and curve in groin,
> And nape set far behind the head:
> Such were the dogs that Fingal bred.

Gesner, in his history of quadrupeds, published in 1560, gives drawings of three species of Scottish dogs, which, he informs us, were furnished him by Henry St. Clair, dean of Glasgow.

These drawings are said to represent the three different

---

* I am informed from Scotland, that a tradition still prevails among the Highlanders, of a much larger species of deer than the present having formerly existed in their hills, which they called "miol." (qu. elk)—*W. S.*

species of dogs mentioned by Boece, in his History of Scotland, published 1526, of which the deer-hound is one. This drawing, though a rudely executed woodcut, is full of character, and coincides with the descriptions which have reached us of this dog.

Of the dog known in Ireland under the name of the Irish greyhound, Holinshed, in his " Description of Ireland and the Irish," written in 1586, has the following notice,—" They are not without wolves, and greyhounds to hunt them, bigger of bone and lim than a colt ;" and, in a frontispiece to Sir James Ware's "History of Ireland," an allegorical representation is given of a passage from the venerable Bede, in which two dogs are introduced, bearing so strong a resemblance to that given by Gesner, as to leave no doubt that they are the same species.

The mastiff and the greyhound both appear, from the old Welsh laws, to have been used from a very early period by that people, and were termed by them, the former Gellgi, and the latter Milgi, which latter is evidently the same word with the appellation of Miol chù, given by the Highlanders and Irish to the deer-hound.

Of the mode of hunting and using these dogs, we have descriptions by William Barclay, as far back as 1563, by Taylor, the water poet, and by others.

The term *Irish* is applied to the Highland dogs, as every thing Celtic (not excepting the language) was designated in England, probably in consequence of Ireland being, at that period, better known to the English than Scotland. This is, however, a proof of the similarity of the dogs, and also that they were not then in use in England in the same perfection. Nor is this supposition inconsistent with the account given by Sir John Nicol, of Queen Elizabeth's amusements at Cowdrey Park, in 1595,—" Then rode her Grace to Cowdrey to dinner, and about six of the clock in the evening, sawe sixteen bucks pulled down with greyhounds in a laund,"—since it will be observed, from the use of the term "bucks," that these deer were fallow ; and, probably, the course was paled in, as appears to have been usual on such occasions, from a minute account by the translator of the " Noble Art of Venerie and Hunting," published in London in 1811.

18

Of the courage of the ancient deer-hound there can be little doubt, from the nature of the game for which he was used, but if any proof were wanting, an incident mentioned by Evelyn in his Diary, 1670, when present at a bull fight in the bear garden, is conclusive. He says, "The bulls (meaning the bull dogs) did exceeding well, but the *Irish wolf dog* exceeded, which was a tall *greyhound*, a stately creature indeed, who beat a cruele mastiff."

Here then is further proof that the Irish wolf dog was a greyhound, and there can be little doubt that it is the same dog that we find mentioned under the name of "the Irish Greyhound."

On comparison, therefore, of the descriptions given of the Vertraha of Nemesian, the English greyhound of the 15th century, the Irish wolf dog, and the Highland deer-hound, we find a strong similarity; and when it is recollected, that the game for which they were all used was the same, and that the term *miol chù* was the one generally used for this species of dog over a great portion of the country, we have strong reasons to conclude that they were one and the same kind, the more particularly as we find the Irish wolf dog described as a greyhound, and the Highland deer-hound as an Irish greyhound; and find that the drawings which have reached us of the Scotch and Irish dogs, bear so strong a resemblance to each other.

From the above authorities, it is obvious that this race of dogs has been known in this country for many centuries, and for a greater period of time than any other sort; indeed, it is the opinion of most naturalists, and, among others, Buffon, that they are an original race, and natives of Britain. On this subject he has the following remarks:—"The Irish greyhounds are of a very ancient race, and still exist (though their number is small) in their original climate: they were called by the ancients, dogs of Epirus, and Albanian dogs. Pliny has narrated, in the most elegant and energetic terms, a combat between one of these dogs, first with a lion, and then with an elephant; they are much larger than the mastiff. In France they are so rare, that I never saw above one of them, which appeared when sitting to be about five feet high, and resembled in figure the Danish dog, but greatly exceeded him in stature. He was totally white, and of a mild and peaceable disposition."

In corroboration of Buffon's theory, that the dogs of Epirus and Albania are the same with the Highland deerhound, it may be remarked as not a little singular, that the dogs at present in use in the mountains of Macedonia, for the purpose of deer-coursing, are similar in figure, colour, disposition, and in the texture of their hair, to those used in this country. They are only to be found in the possession of the nobility, and are with them also exceedingly rare.*

The exact size to which the deer-hound once attained in this country it is now difficult, from the contradictory accounts that have reached us, to determine.

Buffon, as we have already seen, informs us, that the only one he ever saw was much larger than a mastiff, and when sitting was about five feet high.

Goldsmith, in his account of the species of dog known in Ireland in his time, under the name of "Irish wolf-dog," represents him as being rather kept for show than for use, there being neither wolves nor any other formidable beast of prey in Ireland that seem to require so powerful an antagonist.

Judging also from the drawing of Lord Altamount's dogs, given by Mr. Lambert, and from the measurements taken by him in 1790, it is evident that these wolf-dogs, as they are called, bore no resemblance whatever to the Irish greyhound, as described by Holinshed, with which also they hunted wolves, as is apparent from their broad pendulous ears, hanging lips, hollow backs, heavy bodies, smooth hair, straight hocks, drooping tails, and party colour; but were, in all probability, a remnant of the old Irish blood-hound, which was frequently used for tracking wolves, and which at a later period might have been mistaken for a species then in that country nearly, if not altogether extinct.

---

* My friend, Mr. Skene, is possessed of an ancient and curious map of the world, in which the ert, or elk, is represented as characterising the Transylvanian Forest, and near it is a representation of "Canes fortiores," or the great Albanian dog, which these northern tribes are reported to have used to drag their carriages, as well as to hunt the bear, wolf, and elk. The animal given as the elk in the map is represented with very broad palmated horns, more like those of the moose deer, or the extinct *Cervus euryceros*, whose remains are found in the bogs of Ireland and the Isle of Man, than the true elk. This serves to connect the miol-chù of Ireland and the Highlands still more closely with the Albanian deer-dog.—*W. S.*

To these vague accounts, however, little weight can be attached, and the only real criterion by which we can form a notion of the perfection to which this breed formerly attained, is from the small remnant that we now possess.

In Ireland at the present day (we speak from the most accurate information) not a vestige of this breed is to be met with.

To England the same remarks may be applied. In Wales some of this breed may still exist, although no evidence of the fact has reached us. In Scotland (from a perfect knowledge of every specimen of the breed) we know that very few—perhaps, not above a dozen—pure deer-hounds are to be met with.

It is difficult, without a great variety of measurements, to determine the exact size of a dog, or to give an accurate idea of its proportions; though a good general idea may be formed, by giving the height at the shoulder, as measured with a slide, the girth round the chest, and the weight of the dog, together with a few descriptive remarks regarding him.

Applying, therefore, the above rules to such of this race as we have seen, and allowing for the degeneracy which must have taken place in this breed throughout the country (arising from diminution in number, neglect in crossing, selection, and feeding), these dogs may probably have, at a remoter period, averaged in height thirty inches, in girth thirty-four inches, and in weight 100 lbs.

Notwithstanding the degeneracy above alluded to, none of the canine race present at this day such a combination of qualities as the Highland deerhound,—speed, strength, size, endurance, courage, perseverance, sagacity, docility, elegance, and dignity; all these qualities are possessed by this dog in a very high degree, and all of them (with the exception of the two latter) are called eminently into exertion in pursuit of the game, for which he is so well calculated. Every attempt to improve this race by a cross with any other species has utterly failed. Such has been the result of the attempts made with the bull-dog, the blood-hound, and the Pyrenean wolf-dog; by the cross with the bull-dog courage was gained, but speed, strength, weight, and that roughness which is necessary for the protection of the feet in a rocky mountainous country, was lost. In the cross with the blood-hound no quality was gained but that of smell, while the speed and size were diminished; and with the Pyrenean wolf-dog, though weight was in some cases gained, yet this was of no avail, as speed and courage were both lost.

All these crosses were found totally unfit for the purpose of deer *coursing*, as was effectually proved by the late Glengarry, who made many attempts to perpetuate this sport. Of the cross with the blood-hound was Sir Walter Scott's dog, bred and presented to him by Glengarry.

The finest, I believe, and apparently the purest specimens of the deerhound now to be met with, are those in the possession of Captain M'Neill, the younger, of Colonsay, of which he has in particular two dogs, Buskar and Bran, and two bitches, Runa and Cavack.

These dogs, though all more or less related to each other, vary somewhat in colour, two being of a pale yellow, and two of a sandy red; and vary also in the length and quality of the hair.

There is one peculiarity common to all, viz., that the tips of their ears, eyes, and muzzles, are black, and that in all other parts they are each of one uniform colour, a never-failing accompaniment of purity of breed.

In their running points they bear a great similarity to a well-bred greyhound; and, though somewhat coarser, are supposed (from the trials which have been made) to be quite as swift. Their principal difference in shape from the common greyhound consists in a greater height of shoulder, thickness of neck, size of head and muzzle, and coarseness of bone. They are much more sagacious than the common greyhound, and in disposition are more playful and attached, but much bolder and fiercer when roused.

The following are the dimensions of Buskar,[*] taken in August, 1836:—

---

[*] The principal dog in Mr. Edwin Landseer's beautiful vignette, opposite to the frontispiece of this work, is taken from a sketch of this celebrated animal, but does not, I think, give the idea of quite so much bone and muscle as belongs to the original.—*W.S.*

Height at shoulder,	. .	28 inches.
Girth of chest, . . .	. .	32 ,,
Weight in running condition, .	. .	85 lbs.

This dog is of a pale yellow, and appears to be remarkably pure in his breeding, not only from his shape and colour, but from the strength and wiry elasticity of his hair, which by Highlanders is thought to be a criterion of breeding.

Though the dogs now described are of a yellow or reddish colour, yet there are in the districts of Badenoch and Lochaber, some of a dark grey, which are considered pure; indeed it is believed that this was at one time the prevailing colour in the Highlands of Scotland. Besides the difference of colour, there seems to be a decided difference in the texture of the hair between the yellow and grey dog; that of the grey dog being much softer and more woolly. The latter also seem to be less lively, and do not exhibit such a development of muscle, particularly on the back and loins, and have a tendency to eat hams.

There is a striking peculiarity in the deerhound, viz., the difference in size betwixt the male and female, which is more remarkable than in any of the other varieties of the canine race.

The following are the dimensions of a full-grown stag taken from actual measurement:—

	Ft.	In.
Height at shoulder, . . .	3	11¼
Girth at shoulder, ...	4	7¼
Height from top of head to the fore-foot, .	5	6
Length of antler, ...	2	6
Extreme height from the top of the antlers to the ground, ...	7	10
Weight as he fell, 308 lbs.		

When we consider the above measurements, it is not a matter of surprise that few dogs, if any, should be found, who are capable, single handed, of pulling down an animal of such size, strength, and activity.

Deer-coursing, the noblest of all the Highland sports, has long been a favourite amusement with the inhabitants of the north and west of Scotland; and though fallen into disuse of late years, it is still practised in some parts of the country. For the following account of the mode in which it is now practised we are indebted to one of the few sportsmen, who have had the good fortune to enjoy (of late years at least) the pleasures of this exciting sport.

It was on the evening of the 11th of August, 1835, that a party, consisting of six sportsmen, a boat's crew of seven men, with piper, deer-stalker, and two deerhounds, set out from Colonsay, and landed on a beach on the north and precipitous coast of Captain M'Neill's property in the island of Jura, and having clambered up a broken and rocky bank to the foot of a precipice which overhung the sea, they entered by a gradual slope into a spacious and picturesque cave, the mouth of which could not be discovered from below. Their first care was to kindle a fire, the smoke of which rose in a straight column to the roof, and crept along almost imperceptibly to the opening, from which it made its escape. Preparations were then made for a repast, one of the sailors officiating as cook. His knowledge of the science of gastronomy was not great, but with the aid of the King of Oude, etc., etc., he contrived to set before us a dish which would have done honour to a greater artiste, and to which our good appetites enabled us to do ample justice. Our repast concluded with the never-failing accompaniment of whiskey toddy; after which, all were anxious for repose, that they might be on the alert by break of day.

By the side of the fire a couch was spread of dried ferns and heather, such as fair Ellen provided for King James; but though our attendant was neither young nor of the fair sex, we had the advantage over royalty in one respect, being provided with a good stock of blankets, a comfort not at all to be despised in such a situation.

At a little distance the sails were spread for the boatmen, and further off, in a recess of the cave, the dogs were fastened to a stone large enough to have secured even those of Fingal, where a bed of dry ferns was laid for them.

The different picturesque groups, and the deep gloom of the cavern, illuminated only by the fitful blaze of the wood fire, presented a subject worthy the study of a Rembrandt, while the sullen roar of the waves as they dashed against the rocks below, and were re-echoed in the cave, gave a wildness and grandeur to the scene, that was romantic and impressive.

Having betaken ourselves to our resting place, sleep gradually stole over the whole party, and it was only at break of day that the lively air of "Hey Johnny Cope," blown from the pipes of **Duncan** M'Carmick, aroused us from our slumbers.

In a moment each sprang from his **couch of** heather; and not forgetting to give instructions for the preparation of breakfast (and in particular, that **the à la** blaze should be again put in requisition), we descended to a **stream**, which runs through the valley at the foot of **the cave, to perform** our ablutions, and having refreshed ourselves with a dip in the sea, returned to breakfast even at that early hour, with no want of appetite. Our morning meal was soon over; Buskar and Bran were got in readiness, and the whole party issued forth **full** of expectation; indeed, so eager were the dogs, that though they had not tasted **food** from **the forenoon** of the previous **day,** they would not look at the cake which was offered them, and Buskar, when pressed, **at length took the cake in his teeth, and** impatiently threw it from him.

From the **lofty situation of the mouth** of the cave, the **view was most extensive and** picturesque. To the right the Atlantic **rolled beneath us, from** whose bosom the sun had just emerged; **before us** lay a wide extended heath, **from** which the mists of the morning had withdrawn, though they still **concealed** from our view the picturesque tops of the mountains by which it was **bounded.** A beautiful valley stretched to the left, divided down the centre by **a deep** ravine, through which **a** mountain stream flowed and emptied itself into the sea immediately below **us,** while over our heads hung a precipitous **ridge of rocks.** All was, as Johnson has expressed it, "rudeness, silence, and solitude." **There was no trace of the** habitation of **man;** not **a sound was to be heard,** except the murmur of the waters, **and** occasionally the wild note of some sea bird as it flitted from rock to rock.

Before leaving our commanding situation, it was deemed

prudent to scrutinise narrowly with our telescopes the ground before us, particularly those beds of fern, so frequent in these moors, in which the stags, having pastured all night, generally secrete themselves on the approach of day, leaving nothing visible but their light grey heads and horns, which, without the aid of a glass, it is impossible to distinguish.

Having satisfied ourselves that there were none within our view, the next point to be considered was the direction of the wind, and the nature of the ground through which we were to pass.

The direction in which we should proceed being agreed upon, Finlay (than whom a better deer-stalker never trod the heath) set out about fifty yards in advance, provided with a telescope; while the rest of the party followed slowly and silently with the dogs in slips. We had thus proceeded up a rocky glen for some miles, gradually ascending from the sea, when the stalker descried (without the aid of his glass) a stag about a mile off. He immediately prostrated himself on the ground, and in a second the whole party lay flat on the heath; for even at that great distance we might have been discovered by the deer. Finlay then returned, crawling along the ground, to the spot where we were lying, and directed us to creep back for a short distance until we were out of sight. As yet, the rest of the party had seen nothing of the stag, and although the stalker pointed steadily in the direction in which he was, not one of the party could discover him with the naked eye; but Buskar, who had hitherto followed quietly, now commenced a low whining noise, and with ears erect, gazed steadily at the spot where the deer was lying. On taking the glass, we were soon satisfied of the correctness of the stalker's vision, for we could distinctly perceive a fine stag lying on the side of the valley to our left, quietly chewing the cud, and looking round in all directions. We immediately retreated, and following our guide, got into the channel of a mountain stream, which (though the stag was in a situation that commanded the greater part of the valley) enabled us, from its depth and windings, to approach towards him until we should be screened by some intervening rocks.

We then left the channel of the stream, and finding that we could proceed no further in that direction without being observed or scented by the deer, whose power of smell is most acute, we turned to the left, and keeping the lowest ground, proceeded some way up the side of the valley on which he lay, when Finlay informed us that we should soon be again in sight; and that in order to keep ourselves concealed, it was necessary to throw ourselves on our faces, and creep through some rushes that lay before us. This we did, following each other in a line, and closely observing the motions of our guide, for a distance of 100 yards, until a rising ground intervening between us and the deer, permitted us to regain an upright posture. Having gained this point, Finlay thought it necessary to take another view of the deer, in case he might have changed his position, and thus, perhaps, be brought into sight of us when we least expected it: it was proper also to ascertain whether or not there were any deer in his neighbourhood, who might be disturbed by our approach, and communicate their alarm to him. For this purpose, unbonneted, his hair having been cut close for the occasion, he slowly ascended the rising ground betwixt us and the deer, looking at every step to the right and to the left, and raising himself as if by inches, with his head thrown back so as to bring his eyes to as high a level as possible. Having, at length, caught a view of the deer's horns, he satisfied himself that he had not moved, and having sunk down as gradually and slowly as he rose, that he might not by any sudden movement attract the attention of the deer, he returned to us, and again led the way; and, after performing a very considerable circuit, moving sometimes forwards, and sometimes backwards, we at length arrived at the back of a hillock, on the opposite side of which, he informed us in a whisper, that the deer was lying, and that, from the spot where we then stood, he was not distant 100 yards. Most of the party seemed inclined to doubt this information, for they verily believed that the deer was at least half a mile to the right; but Finlay's organ of locality was so visibly and strongly developed, and his practice in deer-stalking so great, that the doubts of the party were suppressed, if not altogether removed. Buskar, however, soon put the matter beyond question, for raising his head, he bounded forwards, and almost escaped from the person who held him. No time was to be lost: the whole party immediately moved forward in silent and breathless expectation, with the dogs in front, straining in the slips; and on our reaching the top of the hillock, we got a full view of the noble stag, who, having heard our footsteps, had sprung to his legs, and was staring us full in the face, at the distance of about sixty yards.

The dogs were slipped; a general halloo burst from the whole party, and the stag wheeling round, set off at full speed, with Buskar and Bran straining after him.

The brown figure of the deer, with his noble antlers laid back, contrasted with the light colour of the dogs stretching along the dark heath, presented one of the most exciting scenes that it is possible to imagine.

The deer's first attempt was to gain some rising ground to the left of the spot where we stood, and rather behind us; but, being closely pursued by the dogs, he soon found that his only safety was in speed; and (as a deer does not run well up hill, nor like a roe, straight down hill), on the dogs approaching him, he turned, and almost retraced his footsteps, taking, however, a steeper line of descent than the one by which he ascended. Here the chase became most interesting; the dogs pressed him hard, and the deer, getting confused, found himself suddenly on the brink of a small precipice of about fourteen feet in height, from the bottom of which there sloped a rugged mass of stones. He paused for a moment, as if afraid to take the leap, but the dogs were so close that he had no alternative.

At this time the party were not above 150 yards distant, and most anxiously waited the result, fearing from the ruggedness of the ground below that the deer would not survive the leap. They were, however, soon relieved from their anxiety; for though he took the leap, he did so more cunningly than gallantly, dropping himself in the most singular manner, so that his hind legs first reached the broken rocks below: nor were the dogs long in following him; Buskar sprang first, and, extraordinary to relate, did not lose his legs; Bran followed, and, on reaching the ground, performed a complete somerset; he soon, however, recovered his legs; and the chase was continued in an oblique direction down the side of a most rugged and rocky brae, the deer apparently more fresh and nimble than ever, jumping through the rocks like a goat, and the dogs well up, though occasionally receiving the most fearful falls.

From the high position in which we were placed, the chase was visible for nearly half-a-mile. When some rising round intercepted our view, we made with all speed for a higher point, and, on reaching it, we could perceive that the dogs, having got upon smooth ground, had gained on the deer, who was still going at speed, and were close up with him. Bran was then leading, and in a few seconds was at his heels, and immediately seized his hock with such violence of grasp, as seemed in a great measure to paralyse the limb, for the deer's speed was immediately checked. Buskar was not far behind, for soon afterwards passing Bran, he seized the deer by the neck. Notwithstanding the weight of the two dogs which were hanging to him, having the assistance of the slope of the ground, he continued dragging them along at a most extraordinary rate (in defiance of their utmost exertions to detain him), and succeeded more than once in kicking Bran off. But he became at length exhausted; the dogs succeeded in pulling him down, and, though he made several attempts to rise, he never completely regained his legs.

On coming up, we found him perfectly dead, with the joints of both his fore-legs dislocated at the knee, his throat perforated, and his chest and flanks much lacerated.

As the ground was perfectly smooth for a considerable distance round the place where he fell, and not in any degree swampy, it is difficult to account for the dislocation of his knees, unless it happened during his struggles to rise. Buskar was perfectly exhausted, and had lain down, shaking from head to foot, much like a broken-down horse; but on our approaching the deer, he rose, walked round him with a determined growl, and would scarcely permit us to approach him. He had not, however, received any cut or injury; while Bran showed several bruises, nearly a square inch having been taken off the front of his fore-leg, so that the bone was visible, and a piece of burnt heather had passed quite through his foot.

Nothing could exceed the determined courage displayed by both dogs, particularly by Buskar, throughout the chase, and especially in preserving his hold, though dragged by the deer in a most violent manner. This, however, is but one of the many feats of this fine dog. He was pupped in autumn, 1832, and, before he was a year old, killed a full-grown hind single-handed.

The deer was carried to the nearest stream, which was at at no great distance, for the purpose of being washed; which ceremony, being performed, we sat down to lunch in great spirits with the result of our day's sport; and having concluded with a bumper to the success of our next chase, our only remaining duty was to convey our deer to the cave, a distance of two miles, by the nearest way through the moor. The stag weighed upwards of seventeen stone, but our stout Highlanders, by relieving each other alternately, carried it this distance in the space of little more than an hour. We then took boat, and in a couple of hours were again on shore in Colonsay.

The speed of a deer may be estimated as nearly equal to that of a hare, though in coursing the latter, from its turnings and windings, more speed is probably required than in coursing the former; but, on the other hand, if a dog is in any degree blown when he reaches a deer, he cannot preserve his hold, nor recover it if it is once lost; indeed, it is only from his superior speed and bottom that a dog can continue to preserve his hold, and thus by degrees to exhaust the deer, till at length he is enabled to pull him down.

This great power of endurance is only to be found in a thorough-bred greyhound; for even though a cross-bred dog might succeed in fastening on a deer, he seldom has the speed or endurance necessary for preserving his hold; and should he receive a fall, will, in all probability, suffer much more than a greyhound, whose elasticity of form is better calculated to endure such shocks.

Perhaps the greatest advantage possessed by superiority

of speed is, that the dog runs less risk of injury; for so long as the deer has the power of movement, he will not turn round, or attempt to defend himself with his horns, but endeavours to fly from his pursuers until they have fastened on him, and are enabled, by seizing some vital part, to pull him down; whereas a cross-bred dog, who has not sufficient speed for a deer, and succeeds only in running him down by the nose (and that after a long chase), finds the deer at bay, with his back against some rock; in this situation, no dog can possibly attack a deer with the slightest chance of success. In fact, so skilfully does he use his horns in defence, and with such fury does he rush upon the dogs, that none can get to close quarters with him without the certainty of instant death. In this position, indeed, he could, without difficulty, destroy a whole pack. When running obliquely down a hill (which is a deer's forte), no dog can equal him, particularly if the ground is rough and stony; and, in such a situation, a dog, without great roughness of feet, is perfectly useless. It is therefore advisable not to let loose a dog at a deer in a lofty situation, as the ground is generally most rugged near the tops of the hills, and the dogs run a great risk of being injured. On the other hand, in low and level grounds, a dog is an overmatch for a deer in speed, and, as the deer generally attempts to make for the high grounds for security, and is a bad runner up hill, the dog has a decided advantage when slipped at a deer in such a situation.

It must be a subject of regret to the sportsman and naturalist that this noble race of dogs is fast dying away, and will, in the course of a few years, inevitably become extinct, unless some extraordinary exertions are made on the part of those who are still possessed of the few that remain.

Should they once be lost, it is difficult to imagine how any race of dogs can again be produced possessing such a combination of qualities.

FORESTS OF SCOTLAND.

THE SUTHERLAND FORESTS.

[*Chiefly from the Communication of* Mr. Taylor.]

The bounds of the Sutherland forests have been much limited of late years, as a necessary consequence of the improved system of sheep-farming which has universally taken place.

Recurring to former days, the two largest and most important of these forests were the Dirrie-Chatt and the Dirrie-More.

The Dirrie-Chatt, or the forest of Sutherland proper, was, according to its ancient boundaries, a very extensive, varied, and celebrated hunting forest; it extended parallel with the eastern coast of Sutherland, and at a short distance from it, and it included the interior parts of the county towards the west and north, until it joined the Dirrie-More, and thence passed in an easterly direction to Caithness, along the old boundary with Strathnaver.

An elevated tract of ground from Ben-Leod, near the confines of Assynt, runs eastward through the centre of the county of Sutherland to Ben-Griam-Beg, and from thence to the heights of Knockfin, at the march between Sutherland and Caithness; and this natural feature of the interior of the country was, with some slight variations, the northern boundary of the Dirrie-Chatt. This central ridge is marked by mountains, with intervals of table land; and the rains that fall on these high and continuous summits, find their way in streams or torrents in different directions to the east, or to the north coasts of the county: part of these waters form the sources of the rivers that pass into the

German Ocean; and the remainder, the sources of others that enter the Ocean, along the north coast of Sutherland, from the river Hope to the confines of Caithness.

These were considered the ancient boundaries, but others somewhat different were adjusted, when Lord Reay was proprietor of Edrachilles.

Ben Klibreck, which rises to an elevation of 3,200 feet, is situated to the north of this ridge, and forms the dominant object in the scenery. Although one of its shoulders separates Loch Naver from the romantic and lonely waters of Loch Veallach and Loch Corr, part of the grounds on the east side of these two lakes, as well as the wild solitudes between them and the mountain, were not comprehended within the Dirrie-Chatt, because the waters of Corrie-na-farn, and of the two lochs, all fall into the river Naver, by the river Meallart. Ben Klibreck, and the romantic features around it, formed of themselves a separate and celebrated forest.

From the southern base of Ben Klibreck, above Strath Baggestie, the boundary of the Dirrie-Chatt proceeded to a place called Garslary, and passing close to Craigna-lochan, kept along the eastern side of Loch Veallach and Loch Corr, including within the Ben Ormin forest the finely wooded side of Loch Orr, called Tugarve, one of the most favourite harbours for deer in that romantic district, covered, as it is, with thriving natural birch wood, for an extent of about six miles. Corrie-na-farn, and an outskirt of Truderscaig, originally followed the Klibreck forest, although the Ben Ormin foresters hunted without opposition on the shores of Loch Corr.

From the north end of Loch Corr, the boundary of the Dirrie-Chatt followed the river Meallart, which flows from that loch; making a sharp angle at Truderscaig, it then proceeded to the north of the loch of that name, including Holmaderry, the whole of which is within the Ben Ormin forest; from thence it went on in a direction nearly parallel to the river Naver, as far as the Ravigil rocks. Within these bounds is the celebrated mountain Ben Ormin, in former times the spot selected and preserved for the exclusive hunting of the earls of Sutherland.

Ben Ormin is 2,500 feet high, and between its lumpish shoulders, called Craig-More and Craig-Dhu, lies what was formerly one of the most celebrated deer passes in the north of Scotland. From the Ravigil rocks, the boundary passed into Ben Maedie, including the whole of Ben-Griam-More, and, continuing along the summit of Ben-Griam-Beg, proceeded towards the Beallach-More, leading into Caithness at the height of Knockfin. The hilly ridge that separates Caithness from Sutherland is strongly defined, and forms the eastern boundary of the Dirrie-Chatt, from the heights of Knockfin to the bold headland of the Ord.

From the head of the Ord, the southern line of march of the Dirrie-Chatt followed the mountain belt that skirts the low cultivated land along the coast as far as Craig-More, near the mouth of the river Fleet; and thence it proceeded westward by the side of the valley of the Fleet, and along the hilly ground north of Rhine, and of Lairg Church, as far as Loch Shin; passing still westward along the whole extent of this lake to Corry-Kinloch, and thence to Ben-Leod, where the description of the boundaries of Dirrie-Chatt commenced.

Such were the ancient boundaries of this extensive forest, which stretched from Ben-Leod to the Ord of Caithness, a distance of about fifty miles. Its breadth varied from ten to thirty miles. It comprehended within its limits the following five minor forests, which had their separate annals and traditions:—1. The forest of Ben-Griam; 2. the forest of Sledale; 3. the forest of Ben-Horn; 4. the forest of Ben-Ormin; 5. the forest of Ben-Hee.

The great forest of Dirrie-More differs essentially in its scenery from all the other forests in Scotland; less in extent than the Dirrie-Chatt which adjoins it, all its parts are broken and disjointed in a singularly wild and abrupt manner; and so uniform is this character, that any one section of the interior solitudes of the Dirrie-More would afford a correct counterpart of all the other features of this wilderness of mountains.

Rocky and precipitous masses, separated by ballochs or narrow passes; deep and desolate glens, with vast masses of mountain wrecks resting their bulk on the level; streams oozing through beds of moss; torrents rushing down the steep ravines; black lakes, highland tarns, and deep morasses:—these are, in comprehensive terms, the characteristic objects that force themselves into notice throughout the extensive range of the Dirrie-More.

Every part of this forest is destitute of wood, except the west side of Ben-Hope, the sides of Stack, and the shores of Loch-More, which are partly covered with brushwood. It was not thus, however, in former times. The boundaries of the Dirrie-More extended from Ben-Leod to the head of Glen-Dhu; thence to the head of Loch-Laxford, the head of Loch-Inchard, and by the Gualin, and the deep valley beyond it, to the head of the bay of Durness, and then on by the balloch leading to Loch-Eriboll.

The east side of Loch-Eriboll, with Ben-Hutig and the Moin, as far as Strathmelness, formed part of the forest; and from the head of the bay of Tongue the boundary went by Loch-Loyal, including Ben-Loyal, and then turned westward to the end of Loch-Maedie; from whence it proceeded near the foot of the high ground to the westward, until it reached Ben-Hee, and thence, by the march of Ben-Hee forest, it passed by Loch-Merkland to Ben-Leod.

The extreme length of this range from north to south is about thirty miles, and its general breadth is about twenty; but near both extremities it does not exceed ten miles. Several mountains stand dominant within the above boundaries, and give their names to three forests, which are included within the general range, although they had distinct divisions, and were under the charge of separate foresters. The names of these forests are,—1. The forest of Ben-Hope; 2. the forest of Fionaven; 3. the forest of Arkle and Stack. The altitude of these mountains, from which the above forests derive their names, will give some idea of the character of the country. Ben-Hope is 3,061 feet high; Fionaven, 3,015; Ben-Spionnue, in the same forest, 2,566. The mountains of Arkle and Stack I have no measure of, but believe they are of no great height.

There are three minor detached forests in Sutherland,

which are not included in the great ones of the Dirrie-Chatt and the Dirrie-More,—1. The Parph; 2. the forest of Klibreck; 3. the Dirrie-Meanach.

The number of deer that wander over the vast forests of Sutherland cannot well be ascertained. About thirty years ago an opinion prevailed that it amounted to 3000. The introduction of sheep farms, and other causes, have materially lessened that number, if, indeed, it was a correct one. So that the harts, hinds, and calves, of all ages, taken collectively, do not probably, at present, exceed the number of 1,500. The calculations of the foresters would lessen that number, and the statements of the shepherds would increase it, their respective interests being diametrically opposite.

Hunts were occasionally upon a grand scale, in this as well as in other forests in Scotland, when the deer were collected by scouts, and driven to certain passes. One of these was along the side of Craigmore, one of the most prominent summits of Ben-Ormin, where there is a station still, called "The Earl's Seat," and farther on there is another, called "Angus Baillie's Seat," having been selected by a forester of that name. There are also the remains of several ancient hunting lodges, which were chiefly constructed on the islands in the freshwater lakes.

There seem to have been two modes of killing deer in the Sutherland district, quite peculiar to the country— one was the erection of an enclosure, called *Garruna-bhiu* (the deer dikes): it was formed of two opposite rough stone walls, about a quarter of a mile in length, and 100 yards apart at one end, this distance being gradually contracted to a narrow opening at the other. The deer having been driven in at the wide end in numbers, could not get into the moor at the narrow extremity without great delay, and thus became an easy prey to the sportsmen. The other method alluded to was formerly practised at two extreme points of the Sutherland forests. A strong force of men collected them in herds near the sea-coast, urged them forwards, and, at length, forced them down the cliffs and crags, and drove them into the water. Boats were concealed amid the rocks, which were put in motion at the proper time, and the deer were attacked with such weapons as were then in use, for I speak of a period previous to the introduction of fire-arms. In this defenceless position of the deer, the slaughter must have been considerable, as it is probable that spearmen and bowmen occasionally leaped from their boats into the waters; the commotion of the waves, the shouting, and the rude mêlée must have exhibited a scene little inferior, in wildness of character, to the Indian mode of hunting on the Red Lake.

Sir Robert Gordon states that this mode of hunting was practised at the Pharo Head (the present Cape Wrath) and adds, "There is another part in Sutherland, in the parish of Loth, called Shletadell (Sledale), where there are red deer; a pleasant place for hunting with grew hounds: here also, sometymes, they drive the deer into the South Sea, and so do kill them." The second place thus alluded to must have been the Ord of Caithness, as it is the only part of Sledale forest where such singular means could be put in execution.

Besides sports of this animating description, the chase of the wolf was followed, in former times, with considerable ardour. Some traditionary notices there are of the destruction of the last wolves seen in Sutherland, consisting of four old ones and some whelps, which were killed about the same time, at three different places, widely distant from each other, and as late as between the years of 1690 and 1700. Indeed, some of these detested prowlers continued to ravage the Northern Highlands, till the disappearance of the pine forests deprived them of retreat and shelter. The last survivors of this rabid race were destroyed at Auchmore, in Assynt, in Halladale, and in Glen-Loth.

The death of the last wolf and her cubs, on the eastern coast of Sutherland, was attended with remarkable circumstances. Some ravages had been committed among the flocks, and the howl had been heard in the dead of night, at a time when it was supposed the villainous race was extinct. The inhabitants turned out in a body, and very carefully scoured the whole country; carefully, but not successfully, for, after a very laborious search, no wolf could be found, and the party broke up.

A few days afterwards a man, by the name of Polson, who resided at Wester Helmsdale, followed up the search, by minutely examining the wild recesses in the neighbourhood of Glen-Loth, which he fancied had not been sufficiently attended to before. He was accompanied only by two young lads, one of them his son, and the other an active herd boy. Polson was an old hunter, and had much experience in tracing and destroying wolves and other predatory animals: forming his own conjectures, he proceeded at once to the wild and rugged ground that surrounds the rocky mountain gully which forms the channel of the burn of Sledale. Here, after a minute investigation, he discovered a narrow fissure in the midst of a confused mass of large fragments of rock, which, upon examination, he had reason to think might lead to a larger opening or cavern below, which the wolf might use as his den. Stones were now thrown down, and other means resorted to, to rouse any animal that might be lurking within. Nothing formidable appearing, the two lads contrived to squeeze themselves through the fissure, that they might examine the interior, whilst Polson kept guard on the outside. The boys descended through the narrow passage into a small cavern, which was evidently a wolf's den, for the ground was covered with bones and horns of animals, feathers, and egg-shells, and the dark space was somewhat enlivened by five or six active wolf cubs. Not a little dubious of the event, the voice of the poor boys came up hollow and anxious from below communicating this intelligence. Polson at once desired them to do their best, and to destroy the cubs. Soon after he heard the feeble howling of the whelps, as they were attacked below, and saw almost at the same time, to his great horror, a full-grown wolf, evidently the dam, raging furiously at the cries of her young, and now close upon the mouth of the cavern, which she had approached unobserved among the rocky inequalities of the place. She attempted to leap down, at one bound, from the spot where she was first seen: in this emergency, Polson instinctively threw himself forward on the wolf, and succeeded in catching a firm hold of the animal's long and bushy tail, just as the fore part of the body was within the narrow entrance of the cavern. He had, unluckily, placed his gun against a rock when aiding the boys in their descent, and could not now reach it. Without apprising the lads below of their imminent peril, the stout hunter kept a firm grip of the wolf's tail, which he wound round his left arm; and although the maddened brute scrambled, and twisted, and strove with all her might, to force herself down to the rescue of her cubs, Polson was just able, with the exertion of all his strength, to keep her from going forward. In the midst of this singular struggle, which passed in silence,—for the wolf was mute, and the hunter, either from the engrossing nature of his exertions or from his unwillingness to alarm the boys, spake not a word at the commencement of the conflict,—his son within the cave, finding the light excluded from above for so long a space, asked in Gaelic, and in an abrupt tone, "Father, what is keeping the light from us?" "If the root of the tail breaks," replied he, "you will soon know that." Before long, however, the man contrived to get hold of his hunting knife and stabbed the wolf in the most vital parts he could reach. The enraged animal now attempted to turn and face her foe, but the hole was too narrow to allow of this; and when Polson saw his danger he squeezed her forward, keeping her jammed in, whilst he repeated his stabs as rapidly as he could, until the animal, being mortally wounded, was easily dragged back and finished.*

These were the last wolves killed in Sutherland, and the den was between Craig-Rhadich and Craig-Voakie, by the narrow Glen of Loth, a place replete with objects connected with traditionary legends. The conflict of Drumderg was

---

* Hogg, the Ettrick Shepherd, has a story somewhat similar to this, which probably he got from the Sutherland drovers; but, in his desire to change the circumstances, and make the tale his own, he has fallen into an error which lessens its probability. He introduces a wild boar as the animal held back by the tail, and not a wolf, although the tail of that animal is proverbially short, and of slender dimensions, and could hardly be grasped firmly by the hand; a sow or boar also invariably roars out most lustily when seized or obstructed, and hence the person in Hogg's cavern must have known from such sounds the cause of obstruction of the light without further inquiry. In Polson's exploit, which was a true one, he had the advantage of grasping the long and rough tail of the wolf; and he wounded an animal that dies without complaining as a sow does, and which, according to Buffon, "never howls under correction like a dog, but defends himself in silence, and dies as hard as he lived."

fought in it. Cairn-Bran stands there, the place where
Fingal's dog, Bran, was buried, and the holy waters of
Tober Massan rise from its mosses, which are supposed to
have cured many diseases. The upright stones of Carriken-
Chligh also stand there, which, as the name denotes, mark
the graves of great men. Nor must we neglect to mention
that stone, of many tons weight, called Clach-macmeas,
hurled to this spot from a distance of some miles by a young
giant of the tender age of two months.

It is well known to all who are aware of the Macpherson
controversy, that poetical notices of Fingal and his warriors
have descended by oral tradition, from an unknown age to
the present generation, amongst persons unable to read or
write, and that such traditions are scattered over the whole
extent of the Highlands. They are recollected only in
fragments, and, even in this broken condition, are known
only to a few of the oldest inhabitants, who imbibed them
in their infancy.

Dermid, says one of these traditions, was beloved by the
wife of one of his friends, but he honourably repelled her
advances. Whilst travelling with Fingal's party through
the forest of Ben-hope, she accidentally splashed herself
with some muddy water; and being piqued at the slight
she had met with, "Behold," said she, "the foul water of
the bog has more spirit than Dermid." This taunt rankled
in his bosom, and made him reckless of danger.

The party soon afterwards roused a wild boar, who was
of such large dimensions, and of so fierce an aspect, that
none of them dared to encounter him singly. Dermid
rushed alone upon the furious brute, and, with the assistance of his dogs, transfixed him with his spear. "Loud
roared the boar in the midst of his rocks and woods," but
Dermid alone had the fame of his slaughter.

In those days it was a test of innocence, if a person sus-
pected of crime, measured with his bare legs and feet, and
with impunity, the bristled back of a dead boar, proceeding
from the tail to the head, against the sharp points of the
bristles. To this ordeal Dermid cheerfully agreed to sub-
mit, to satisfy his friends that he had never injured any of
them. But some invidious person dexterously sprinkled

poison over the bristles, and these having punctured Dermid's skin, whilst measuring the length of the boar, the poison took effect and caused his instant death.

Grana, another female devoted to Dermid, was present, and, in her grief and despair, resolved not to survive her lover; and throwing herself on the point of his sword fell lifeless on his body.

The boar was hurled down the side of Ben-Loyal, and buried close to a mountain stream that runs between two of the scors or pinnacles of Ben-Loyal, still called Ault-Torc (the Burn of the Boar); and the hapless Dermid and his devoted Grana were buried in one grave, and under some trees that grew near the spot. There lies the grey cairn at this present day, still held in reverence by the natives: one person alone ventured to despoil the trees, but misery and misfortune befel him and his family.*

Angus Baillie, of Uppat, was one of the most noted foresters in Sutherland, of whom we have any correct account: he signalised himself in many of the conflicts which were of common occurrence in former times, and particularly in a rocky pass, on the banks of the Blackwater, where he and two of his companions defeated a whole host of Caithness freebooters, with the gun, called Glasnabhean, at that time a novel and dreaded engine of destruction. Baillie was likewise renowned for his dexterity as a bowman and deer-stalker, and thus excited the jealousy of one of the midland foresters who went down to Sutherland to compete with him.

This stranger, being recommended to Baillie's superiors, talked boastingly of his pre-eminence over the Sutherland foresters, either at open feats, or in executing cunning devices for overcoming an opponent. Nay, he said he could kill more deer than Baillie on his own ground, and finished his rhodomontade by saying to his face, "You can no more be compared to me, as a forester, than your old shaggy garron can, as an animal, be compared to the finest antlered stag on the hills."

---

* Another version of the *Bas Dhiarmid*, or the Death of Dermid, has been given by Mr. Grant.

A day was fixed for their competition, and Baillie accompanied the stranger to Ben-Ormin.

He thought himself a stronger man than his blustering visitor, and was determined to vindicate his slighted prowess by making the challenger appear as ridiculous as his boasting had been offensive: little recked he of the consequences.

Now this, our Sutherland man, had no aversion to any awkward trick or gambol, by means of which he could distress his opponent. He was, moreover, learned in traditions, and had heard in what manner a Danish giant was said to have been captured by a man of diminutive size; he, therefore, privately directed one of his men to kill a deer, and to spread the fresh skin of the animal immediately within his bothy, with the inner side uppermost: when Baillie and his challenger arrived at the door, the latter was desired to enter first with many terms of courtesy, and as soon as he stepped upon the slippery surface of the fresh hide, his heels tripped up, and down he came upon his back. Whether or not Angus lent him a helping hand in his tumbling propensity, tradition does not say; but I should rather think he did, for the fall was so heavy, that before he could recover himself, the said Angus Baillie, of Uppat, rolled the skin round him, and bound him in it with some cords he had provided for the purpose. He was sufficiently kind and considerate to leave his head free and exposed, nothing more; and thus Master Bobadil, or Rodomonte if you please, exhibited a pretty fair specimen of an Egyptian mummy, or an Italian bambino.

In what manner the man in durance delivered his sentiments on this touching occasion, tradition does not inform us. But, as he could not walk in this plight, Baillie, with mock humanity, carried him to Dunrobin Castle on his shoulders, where he had previously been taunting and boasting; nay, more, when he approached that fair pile, he was complaisant enough to give him his full honours, by tying a large branching pair of antlers to his shaggy head. The stout porter, having then obtained an interview with his superior, exclaimed, with mock solemnity:—

"A wise man is known by the truth of his prophecy, and here I, the humble garron, am carrying home the horned stag that wandered into strange ground." The stranger was liberated by the gentlemen present, and very prudently marched home with the least possible delay.

Hunting parties in the Sutherland forest were formerly upon an extended scale; there may still be seen the ruins of two very large hunting lodges, of the description which Pennant mentions, in the Strath of Helmsdale, the stones of which now form huge cairns: one of these, near Cayn, appears to have been 108 feet long and 26 feet broad; and the other, which is at Saliscraggy, measures 174 feet in length and 26 in breadth, and is situated on a very pleasant bank of the river Helmsdale, near the old Strath road.

But I have lingered a long while in this romantic country; more, much more could I add did my limits allow of it, for the assistance which has been so obligingly conferred upon me, and which I have acknowledged in the preface to these pages, has been most able and ample; but I must now conclude, adding only in the words of Sir Robert Gordon, " The bodies and mynds of the people of this province (Sutherland) are indued with extraordinarie abilities of nature; they are great hunters and do delyte much in that exercise, which makes them hardened to endure travell and labour."

SOME ACCOUNT OF THE FORESTS AND DEER-HAUNTS IN ROSS-SHIRE.

The extensive estate of Lord Lovat, which ranges westwards from his residence of Beaufort Castle, near Beauly, forms the northern boundary of Inverness-shire, for a long distance dividing it from the county of Ross, and having long been the abode of deer, the appropriation of a large space to their exclusive possession has established a good forest; which the judicious care of the noble proprietor, himself a first-rate shot, and good stalker, will continue to improve. With the boundaries of the Chisholm country I am not acquainted.

The wild country of Strath-Conan, on which we enter to the north, is the commencement of the county of Ross, through the whole of the highland parts of which, with little exception, as well as through the adjoining wilds of Sutherland, it would be difficult to find any district not more or less tenanted by the red deer.

The division of Strath-Conan was long held in unenviable notoriety, as the main stronghold of the illicit distiller in the north of Scotland, a celebrity which it has only lost in recent years. The memory of this may possibly ere long have passed away, but the deer-stalker, who is made aware that the scene of Mr. James Baillie Fraser's tale of the "Highland Smugglers" is laid in the Lovat forest, and adjacent recesses of Strath-Conan, will hardly forget it. To the lovers of romantic fictions, connected with scenes of Nature, and to all those whose spirit is excited by the deep interest which patriotism and tradition have thrown around the "land of the mountain and the flood," these volumes will possess undying charms.

To the west of Strath-Conan lie the two great districts of Applecross* and Gairloch, containing a vast extent of the most rugged mountain scenery. A great part of it is, of course, utterly unimprovable, and, indeed, inaccessible,—thus affording to the deer a secure retreat; while the fine valleys of the west, which lie between the hills, offer abundant pasture. In this part of Ross-shire the deer are abundant, and the thorough knowledge of the sport and unerring rifles of Sir Francis Mackenzie, of Gairloch, and his brothers, have brought in many a noble stag to Flowerdale, the picturesque residence of his family. The singular beauty of this place, which is a small glen, or opening, among the wildest hills, crowded with trees and shrubs of the richest foliage, and decked on one side by the silvery sand and bright waters of the north-west coast, make it, including, as it does, the magnificent Loch-Maree in its neighbourhood, an object well worthy of the traveller's toil.

Deer-stalking is here, however, a truly laborious sport, and requires more than ordinary skill and perseverance. One of the luckiest shots which the writer remembers, was made here, in 1832, by the Honourable Edwin Lascelles,

---

* A separate description of Applecross will be given in the following pages.

who brought **down** a stag, in **full trot, at 312 yards**, being his first essay in the sport.

We next cross the long valley which extends **from** Dingwall, at the **head** of the Firth of Cromarty, by Achnasheen and **Loch-Maree, to the** west coast, and enter upon the heart of Ross-shire, **no part of** which is without deer, **nor** likely to be so, while the old Balnagown, or Freevater forest, which forms its centre, **exists.** Groinyard has its deer, so has Achnasheen, **and the hills near** Loch-Luichart; and the comparatively small forest **of** Fannich, lately a part of the Cromarty estate, **is** perhaps **as sure a place** for the **sport, if kept clear of** sheep, as any in Scotland. Coul,* **the residence** of Sir George Mackenzie, **Bart.,** and Brahan Castle, the residence **of** the family of Seaforth, both **within** seven miles of Dingwall, are seldom without deer in their **woods, and these** noble denizens of the forest may frequently form part **of a** day's sport, at either of these places, with pheasants, partridges, etc., and all the variety of low country shooting. It is almost needless to add, that driving is the **mode in practice** here,—the thick cover precluding stalking, except in rare instances.

Crossing all these large **ranges of hills** we enter the **Balnagown forest, or Freevater,** *i.e.*, the forest of Walter, one of the chiefs of that ancient house.

The mountains in this district are very lofty, and abound on their summits with those broken mossy tracts, where the **experienced deer-stalker looks** with increasing expectation **for his** game. It is much to be regretted that hardly **any** part of this fine forest is kept properly clear **of** sheep; though this is doubtless one cause of the increasing numbers **of the** deer in neighbouring places.

They are accordingly found in Loch-Broom, on the estates of Castle-Leod, Sir Hugh Munro of Foulis, Munro of Novar, and Davidson of Tulloch, in sufficient numbers to make the **pursuit of them a constant sport.**

The estate of **Foulis,** comprising the greater part of the lofty range of Ben-Weavis, should perhaps be more specially

---

*A separate notice of the beautiful possessions of Coul will follow this general account of the deer-haunts in Ross-shire.

mentioned, as capable of being made, by the exclusion of sheep, a sure resort for red deer.

From the Freevater forest, the deer have long since straggled into the large fir woods in Easter Ross, which are in the neighbourhood of Balnagown Castle, and Calrossie, and though they may wander, in many instances, between these woods and their original forest, they have now completely established themselves there, dwelling and feeding amid much interruption from the proximity of population, for which, however, experience has shown that the red deer, in the shelter of his woods, care but little.

To the west of the Freevater forest there remains of Ross, or rather of Cromartyshire, the wild district of Coigach, a part of the Cromarty estate, and the property of the Honourable Mrs. Hay Mackenzie; and the deer-stalker, who loves the sport in perfection, will be glad to learn that the son of this lady has devoted a considerable part of Coigach as a forest for the deer; intending to build a lodge there, at Rhidorach, a situation of much natural beauty.

The isles of Lewis and Harris contain a large number of deer; and in the former, Sir Frederick Johnstone, Bart., who rents the game, has, together with his friends, done great execution; but these deer, I am told, are inferior in size, existing, as they do, in an ungenial and unproductive country, though the climate is fitter perhaps for raising their food than that of man.

A SHORT ACCOUNT OF COUL.

[*Obligingly Communicated by* Sir George Stuart Mackenzie, Bart., *the Proprietor.*]

There are few country residences so favourably situated for sport as Coul. Between breakfast and dinner time you may have amusement with every kind of game, except ptarmigan, which are too remote. I have myself, says the proprietor, brought in a couple of salmon, and a stag has been shot, both within an hour after leaving the house. The increase in the numbers of red and roe deer has been remarkable. Twenty years ago, it was a rare thing to meet with either. It was supposed that the introduction of sheep had driven them away; but though this may have been one great cause, it was neither the sheep nor the shepherds, nor their dogs, that occasioned the extreme scarcity, but the great extent to which poaching was carried—every Highlander having formerly been in possession of a gun of some sort or another. At the residence of Coul there are still preserved some pieces of strange and uncouth appearance, which have at various times been employed on this service. Many of them have Spanish barrels, perhaps relics of the Armada; some are of French construction; and many a gun that had made a noise during the civil wars and rebellions was turned against the stately rangers of the mountains. Nay, in modern times, muskets that had graced the shoulders of volunteers of our own day, by some means or another, had escaped being restored to the armoury of the Tower of London, and remained for efficient ball-practice, as well as for sending showers of small shot amongst grouse and black game. In proportion to the increase of sheep-farming, the numbers of Highland sportsmen were diminished; and to this I attribute the recent very rapid increase of the deer. The attention of English sportsmen was called to them, and the protection since given, has, in some districts, rendered them a nuisance to the farmers.

There are several districts in Ross-shire where deer are stalked; but at Coul they carry on the war by what is called a "tinckel," which, in practice, signifies a drive towards particular spots, or passes. The scenery is very beautiful, and to some points where the guns are usually stationed, the access is so easy that ladies may witness the sport. It is a very fine sight, says Sir George, to see a herd emerge from one part of the wood and scour the open space; sometimes occupying a knoll and reconnoitering, and then dividing into parties, and making for other shelter. Their movements are so exciting, that killing the creatures is not always thought of; and the sportsmen sometimes become so nervous, that they mistake distance, and either miss a near shot, or do not fire at all. Again, a deer has been known to run a-muck along some hundred yards of an opening in the wood, and to receive five balls before he fell. Thus many are lost, which retire to thickets when wounded, where they die.

The hill of Tor-Achilty, close to the beautiful residence of Coul, abounds with deer. It is finely varied, and there is a small lonely lake at its foot; the hills around are covered with birch and oak trees for miles, and deer are found on all of them. Two rivers meet at the base of the hills, and the herds are thus in a manner confined, so that their haunts and ways are perfectly known. Occasionally, though the passes be well watched, not a shot will be fired; and, at other times, much powder and ball is expended in vain. Yet there is always some consolation—the deer were seen,—had a slight change of position been made, a shot would have been got—and so forth.

Fallow deer are in a wild state in the vicinity of the mansion, and they are sometimes seen in the most distant woods. A good many years ago part of the fence of Lord Seaforth's deer-park gave way, and all his lordship's deer escaped to the woods. They are not, however, dispersed to any great distance.

FOREST OF APPLECROSS.

The forest of Applecross lies in Ross-shire, and is comprehended in a circuit of great extent; its boundaries may be traced passing from the north to the east, and so on to the south and south-west, from Inverbain round by Loch Loundy, Beinn Vaan, Cairn-Derg, Coir-nan-a-rog, Coir-na-ba, Coir Scammadale to Solchmore, or Red River, a distance of fifty miles; then again completing the circle, by proceeding from the south-west towards the north and east, and passing from Red River, Beinn-horornaid (or Fairy Bridge), Avy Broch Coir, Bhuochroch, Garry Vaul, Coir Glass, Craikvein, to Loch Gannich, a further distance of forty miles.

The Sanctuary, Coir-Attadale, from north-west to south-east, is six miles long, and there are various warm and fertile corries in all directions, which the deer delight in.

Ault-More, or the Big Burn, is picturesquely wooded; and, as well as Ault-Beg, or the Little Burn, is a favourite retreat of the denizens of the forest. The mosses are everywhere remarkably fertile, and contain innumerable lochs; even the highest hills afford good pasture, and are scattered over with the sea-daisy and other plants. The corries and burn sides are still more rich and verdant.

The numerous lochs in this forest are not only oramental, but valuable for their produce. Loch Coir-Attadale, which empties itself by means of the excellent fishing river of Applecross, is stocked both with loch and sea-trout. Loch Gannich, Loch Na-creig, and Loch Na-long, are likewise amply furnished with the same delicacies, and many of the smaller lakes derive their names from the size and quality of the fish which they contain. The hills in this fine district are strikingly picturesque, and nothing can surpass the beauty of the strath of Applecross.

The deer forest was established about seventy years ago; the quantity of deer it contains at present cannot well be ascertained, but it has been represented to me as very great. They are scattered over their favourite hill sides in such numbers, that when put in motion, and scampering away, they give a character and animation to the scenery quite in keeping with the magnitude of the objects around them.

The anecdotes, which have been obligingly sent me relating to the sports in this forest, are such only as are of usual occurrence. They use the rough stag-hound, or lurcher of the country for wounded deer.

I have received no general account of the weight of these deer; but, judging from the size of others on the western coast, I am inclined to estimate it at a high rate, particularly as it is recorded that Thomas, the first laird of Applecross of that name, killed two stags a few years before his death, that had been destroying the corn a short distance from the mansion-house, whose weight was sixty pounds the quarter.

THE FOREST OF GLENGARRY.

The Glengarry forest is situated in Inverness-shire, and is about seven miles long from east to west. On the north it is bounded by Glen Loing, and on the south by the ridges of the hill.* Part of the ground consists of good pasture, with rich meadow land on the banks of the river; on the northern part there is long heather and reed, and near the top of the ridges, much sweet grass, of which the deer are particularly fond. The whole of this extent has been preserved from sheep for about forty-six years, and is still retained as a forest, generally known by the name of "Eisnich." Stags, however, are not found in it in great numbers, except in the rutting season. The late Glengarry preserved the greater part of this ground as a "Sanctuary," never permitting any one to hunt in it, even in pursuit of a wounded deer; thus, when the game was disturbed on the neighbouring hills, they made towards this spot as their refuge. The pasture being good, the climate comparatively mild, and the snow never lying long on the ground, are circumstances so favourable, that the deer attain to a large size. The late Glengarry killed a hart, which weighed twenty-six stone, and the present proprietor, another of the weight of twenty-four stone five pounds, both weighed after the gralloch had been taken out. The latter deer had previously been wounded in the shoulder by the same gentleman about ten days before the last decisive shot, by which occurrence he was somewhat wasted.

The mode of killing deer at present practised in this forest is such as would naturally be used in any other ground of a similar nature. They are stalked on the hills, and in the lower ground the woods are driven, whilst the passes are occupied by the rifle-men. Formerly there were grand hunts, when the herd was driven into lake Dulachan by a strong *cordon* of men, and the slaughter took place in its waters.

The late Glengarry, amongst other things, was celebrated

---

* By the hill, I believe, is meant the general mountain range which rises from the Strath.

for the excellence of his deer-hounds: who, indeed, has not heard of the remarkable feats of *Hector?* He tried various crosses, particularly with a small blood-hound; and their capacity of following a cold scent is said to have been so wonderful, that one of them actually pursued a wounded deer for the space of three days, the hunters at nightfall stopping at the last distinct impression of the deer's hoof, and covering it with stones; when the stones were removed at daylight, the hound was put upon the scent, and went forward as keenly as ever.*

Many of Glengarry's dogs met the fate common to all high-couraged ones, and were occasionally wounded by the antlers of the stag at bay, or fell over precipices in turning a sharp corner during the heat of the chase.

With what romantic ardour the late Glengarry followed up the exciting amusement of deer-stalking, is well known throughout Scotland. He would go forth in his kilt, and remain on the hills for a week together, sleeping in the open air. When the stag was at bay, he would sometimes have a close engagement with him, using his gun-stock, or skene-dhu, and, though often in peril, was ever successful. Stout-hearted and enthusiastic as he was, nothing could obstruct his course: when his dogs once held a stag at bay in an island in Loch Garry, no boat being at hand, he placed a knife in his handkerchief, which he bound round his head, swam lustily through the waters, and completed his victory. This was wild sport, indeed; but he had an adventurous and a gallant spirit, and was a true son of the mountains.

THE DUKE OF GORDON'S DEER FORESTS.

At page 100 will be found an account of the former possessions of the Earls of Huntly. As some changes of consequence have taken place in latter times, perhaps it may be as well to note the more modern measurement and divisions

---

* It must be borne in mind that a wounded deer would not hurry on unless closely pressed, so that the scent was not so stale as it would appear from this account.

of this wild tract, precisely as I have received them from another quarter: should there be any discrepancy between the two accounts, the changes above mentioned, and the difference between computed and actual measure, will easily account for it.

GLENFESHIE,

in the parish of Kingussie and county of Inverness, is bounded on the south and south-east by the forests of Marr and Atholl; on the west, by the forest of Gaick; and on the south by the estate of Invereshie; by survey in 1770, it contained 13,706 Scots acres. It was let in 1752 to Mr. Macpherson of Invereshie, and continued to be rented by that family until 1812, when it was purchased from the Duke of Gordon by Mr. Macpherson of Invereshie and Ballindalloch. It has been pastured by cattle and sheep since 1752.

GAICK,

in the parish of Kingussie and county of Inverness, is bounded on the south and west by the forest of Atholl, on the east by the forest of Felaar, and the estate of Invereshie, and on the north by the lands of Invertruim, Ruthven, Noid, Phoness, and Glentruim. It contains three lakes stocked with char and large trout, and salmon are occasionally found in them, ascending by the water of Iromie from the Spey. By survey in 1770, it contained 10,777 acres. It was let in 1782 as a sheep-walk to Robert Stewart of Garth for nineteen years. In 1804 it was let to Colonel Gordon of Invertruim, who occupied it as a grazing till 1814, when the Marquis of Huntly got it from his father as a deer forest. In 1830 it was purchased by Mr. Macpherson Grant, of Ballindalloch, from the Gordon trustees, and it is now let to Sir Joseph Radcliffe, Bart., who strictly preserves it as a deer forest, and has an excellent shooting lodge near the centre of the range.

DRUMAUCHTAR,

in the parish of Kingussie and county of Inverness, is bounded on the south by the west forest of Atholl, on the west by the Duke of Atholl's and Sir Neill Menzies's properties, and on the north and east by the lands of Glentruim and Cluny. By survey in 1770, it contained 5,782 Scots acres, exclusive of Beinalder, which forms a part of it, and contains 14,927 acres. It was let for pasture to Lachlan Macpherson in 1773. In 1829 it was purchased from the Gordon trustees, along with the lands of Glentruim, by Major Ewen Macpherson, of the H. E. I. S., and is occupied as a sheep-walk and grouse shooting range. Beinalder is now the property of Ewen Macpherson, Esq. of Cluny, and has recently been let to the Marquis of Abercorn as a deer forest.

GLENMORE,

in the parish of Kincardine and county of Inverness, containing 10,173 acres, was formerly a great pine forest. It is bounded on the south by the forests of Glenavon and Marr. It is used now for pasturage. Cairngorm forms part of this forest.

GLENAVON,

in the parish of Kirkmichael, county of Banff, contains 22,086 Scots acres. Since 1773 it has been occupied as a grazing, but it is said that the Duke of Richmond contemplates restoring it to a deer forest. It adjoins the forest of Marr.

GLENBUILY,

adjoining Glenavon, 3,396 acres.

GLENFIDDICH,

parish of Mortlach, county of Banff, 5,522 acres, is possessed by the Duke of Richmond as a deer forest, and has always been retained as such by the Gordon family.

Of all these ancient forests, the last and Gaick are the only ones now strictly preserved for deer; the others are pastured by black cattle, or sheep, and are therefore only partially stocked with the nobler animals.

THE FOREST OF INVERCAULD.

THE Invercauld forest is situated in the parish of Braemar, and county of Aberdeen. Lord Byron's famed Loch-na-

Garbh* is on the extreme east point, and Bein-a-bour, and Bein-avon guards it on the west and north. The river Dee divides it, flowing from west to east, and its numerous small tributaries afford abundance of the finest water for the animals grazing within the range. The house of Invercauld is nearly in the centre of the sport, and may be said to be surrounded by the forest; as through the spring and winter months the deer may daily be seen browsing about almost within gun-shot of it, and the destruction they do to the numerous plantations shows they are at no great distance during the rest of the year. With a glass they can be viewed at any time from the windows on the hills around. The extreme length of this forest from east to west is eighteen miles; the breadth varies from two to five; it is equal to thirty-four square miles; the circumference is forty-two miles, and it contains 22,186 acres. Within this extent you find every description of ground, from the bold rocky mountains of 4000 feet in height (on which have been found many stones of the topaz and beryl kind), to the table land of the district 1,100 feet above the level of the sea. The pasture varies from the finest natural grasses to the lichen and pure white or grey fog on the summit of the hills; but the heather and ling predominate, and these latter are from time to time renewed by burning. With abundant shelter from the woods and plantations, and such excellent pasture, no situation can be more favourable for the protection of deer. The junction with the Marr Lodge forest on the west increases greatly the value of both; that again joining with the Atholl forest, which latter is contiguous to Gawick, forms altogether a greater extent of connected surface, kept expressly for deer, than is to be found elsewhere in Britain. A glen, joining the Invercauld forest on the east of Loch-na-Garbh, called Glen Gelder, has lately been reserved for deer by Sir Robert Gordon, which, from the increased extent of ground, and protection it affords them, must prove an advantage to both parties.

---

* The height of Loch-na-Garbh, according to the proprietor, is 3,824 feet; that of Beinn-a-bourd 4,039. Beinn-avon has 5,967; and Beinn-a-muich-dui, also in the forest, is represented by him as the highest mountain in Scotland, being 20 feet higher than Ben-Nevis. The wild character of the country may be easily divined from these majestic features.

There is no tradition how long this range has been under deer; it has always been considered part of the Royal Forest of the old Scottish kings*, and there are still the remains of Kindroghit Castle on the Invercauld property, used by Malcolm Canmore as a hunting seat, of which mention is made in the notes to Sir Walter Scott's "Marmion." It does not exceed twenty years since the sheep and other animals were finally cleared off the hills to the west, where it joins the Marr Lodge forest. About 4,200 acres are in wood, the greater part of which, on the east side, called the Ballochbui, consists of indigenous pine, many centuries old, and of great size. There are several hundred hinds which never leave these woods or their vicinity; but it is difficult to estimate the number of deer belonging to Invercauld, as it is constantly fluctuating with a change of wind. In summer, the prevailing west wind frequently takes the greater part of the stags to the Marr Lodge forest; but when the wind returns to the east, or in bad weather, hundreds of stags and hinds immediately come back; and in winter and spring the woods are always full of them. The roe deer at all times abound in these woods. There are no lochs worthy of note in the forest, but there are several in the adjoining grouse ground belonging to Invercauld, the greater part of which is let in different shooting quarters, and all under sheep: when the latter are removed from the hills to winter pastures, the deer, particularly the stags, frequent great part of it until the sheep return in summer. It extends to 112,760 acres, surrounding the forest on the north and south, which, when added to it, makes a total of 134,946 acres, equal to 210 square miles,

---

* There is a letter under the privy seal of James VI., appointing Donald Farquharson of Braemar, keeper of the King's forests of Braemar, Cromar, and Strath-dee, dated 1584, "with power to him, his deputis, and servandis, for quhome he sal be holdin to answer to cause hayne the said wodis, forestis, and mureis; and to serche, seik, tak, and apprehend all and quholsumevir personis hantand, or repairand tharin with bowis, culveringis, nettis, or any uther instrument meit and convenient for the distruction of the deir and the murefowlis; or with aixis, sawis, or any uther instrument or worklume for cutting or destroying of wood; and to tak the same in for thame and intromit thairwith to his awin use; and to present thair personis to the justice, shiref, or any other ordinar juge to be punisheit conforme to the lawis of this realme and generallie, &c.: term and stabill to hold, &c., at Falkland, the 11 day of Jully, the year of God Im V lxxxiv. yeiris."

and is 108 miles in circumference. The names of the principal lochs are Lochbalader, above a mile in length, Loch Kenlader, Loch-nau-eau, famous for its trouts (which are of a red colour resembling those of Loch Leven), Loch Brotichan, and part of Loch Muich, which is the largest, being above two miles long.

The old method of stalking the deer against the wind is the one generally practised in this district: it is not always easily done, owing to the numerous eddies which are met with among the hills; and thus it allows great scope for the knowing forester in displaying his tact, and in bringing the sportsman within shot of his object. In the Ballochbui, the deer are frequently, more especially in cold or windy weather, to be seen within shot of the drives; and both stags and hinds have been often killed by Mr. Farquharson from a carriage or a pony. The deer are seldom driven, never hunted with dogs, unless to bring down a wounded animal. The foresters have small terriers properly trained to keep by them when stalking, and these will track a wounded deer to a great distance without giving tongue, and have been known to find one after two nights and a day had elapsed. They answer every purpose, as they bring the sportsman within a second shot without being perceived, whilst greyhounds, when the object is out of view, cannot follow the scent properly; so that where the ground is stony, or in the woods, they are almost always unsuccessful. Such greyhounds as are in use are descendants of the Glengarry breed, and have been lately crossed with the common foxhound, and also with the bloodhound; but still the foresters prefer the terriers, which are of a very sagacious nature, and were brought originally from Ross-shire.*

The Braemar deer are allowed to be quite different from those of the Atholl forest; they stand higher, and are in general of a greater weight. The stags average from fourteen to sixteen stone gralloched, but occasionally far exceed that, and have weighed (with the inside) as much as twenty-five stone of fourteen pounds to the stone. The hinds seldom

---

* Probably because they made less disturbance in the forest, which, although of a princely extent in point of length, is rather narrow for urging the chase in a cross wind.

exceed eleven stone. The number generally killed at Invercauld in the course of the year is about thirty or forty stags, and twenty hinds.

THE FOREST OF MARR.

The forest of Marr, in the county of Aberdeen, consists of four contiguous glens on the north bank of the Dee, with their various branches and ramifications, viz., Glenquoich, Glenluie, Glendee, and Glenguildy.

The extent must, in a great measure, be guessed at, it never having been regularly surveyed; but as far as can be ascertained from the opinion of those who know the ground well, and have had the best opportunity of judging, it is thought the length of the forest may average fifteen miles, and the breadth eight, which would give an area of about 60,000 acres. The bearing of the extreme length runs nearly east and west. It is bounded on the north by Glenavon in Banffshire, and the hills of Rothiemurcus and Glenfeshie in Inverness-shire; on the west by part of the forest of Atholl and the glen of Baynock; on the south by the river Dee; and on the east by part of the forest of Invercauld. The whole is in the county of Aberdeen. The principal rivers and streams are the Dee, the Quoich, the Luie, and the Guildy. The Quoich branches at the top and runs into the Bechan and Duglin burns; the Luie into the Derry and Luibeg; the Dee into the Garchery and Guirachan; and the Guildy into the Davie and the burn of the Cuirn. The only lakes worth mentioning are Loch Eatechan on the east shoulder of Bein-muirdhui and Loch-nastirtar in Glenguisachan.

The principal mountains beginning on the east are Beinaboard, Beinachuirn, Beinavrear, Beinamean, Cairngorum, Bein Muirdhui, Cairnavaim, Breriach, Cairntoul, Beinavrottan, Cairn-nealler, and Scarrach;—they are all composed of granite; and the general character of their surface is dry and rocky for a considerable way down their sides; but there are many valleys or corries around their bases containing good rich hill pasture; and in the low parts of the different glens are haughs of rich natural grass, which, in Glenluie and Glenquoich, are well sheltered by very extensive tracts of natural pine wood: there is also a considerable proportion of mossy ground interspersed over the whole.

The Glenluie was cleared of sheep and cattle, &c., and turned into a forest upwards of sixty years ago, and the other glens at different and more recent periods.

The number of deer in the forest must vary to a great extent according to circumstances; but it is supposed that there may be a regular stock of about three thousand.

The weight of the best stags may run from fourteen to eighteen stones imperial, and there have been instances of some of the largest weighing twenty stones.

In this forest the deer are for the most part killed by stalking, and getting *quiet shots*, and not by driving them to passes, or coursing them with dogs, except when wounded.*

The breed of deer-hounds chiefly in use is the rough wire-haired Scotch or Irish greyhound.

The present Earl of Fife has tried many spirited experiments by the introduction of different animals into this celebrated forest. He brought over capercalies from the north, and they increased to the number of twelve; but when the place was let, and the birds were removed, they soon died. He has now procured two more old ones; and has succeeded, I am told, in rearing up another brood. The wild boar also was introduced at the advice of the Margrave of Anspach, who was at Marr Lodge for a fortnight, but the experiment did not answer for want of acorns, which are their principal food; if these animals, however, were turned out young, the ant-hills, which abound in the forest, might probably be an efficient substitute. Reindeer were also introduced by his lordship, but they all died, notwithstanding one of them was turned out on the summits, which are covered with dry moss, on which it was supposed they would be able to subsist. In spite of these failures, Lord Fife wished to see if the chamois would live

---

* The little disturbance which this method occasions to the forest keeps the deer from wandering, though the sport is of a less brilliant description.

in his alpine domains, and he imported five of these animals from Switzerland; his late Majesty, however, having expressed a wish to have them at Windsor, they were accordingly sent there, where they produced young ones. A wooden tower was built for them, and they raced up and down it as if they had been amongst their native rocks. They died from having eaten some poisonous herb, so that on all accounts, it is very much to be regretted that they were not sent originally to the Marr Forest.

The remaining trees in Braemar are the last of the Scotch pine-forests: their leaves are of a very dark green as compared with the common Scotch fir.

I wish the communications I have had the honour of receiving from the Earl of Fife had enabled me to give a more detailed account of this magnificent country, and the traditions which belong to it. Unfortunately, I have it not in my power to supply any further information, and shall therefore close this account with an extract from a work of Taylor, the Water Poet, entitled "The Pennylesse Pilgrimage," relating to a great hunt given by the Earl of Marr in 1618.

"I thank my good Lord Erskine," says the poet; "hee commanded that I should alwayes bee lodged in his lodging, the kitchen being always on the side of a banke, many kettles and pots boyling, and many spits turning and winding with great variety of cheere, as venison baked, sodden, rost, and stu'de; beef, mutton, goates, kid, hares, fish, salmon, pigeons, hens, capons, chickens, partridge, moorcoots, heathcocks, caperkillies, and termagents; good ale, sacke, white and claret, tent (or Allegant), and most potent aquœvita.

"All these, and more than these, we had continually in superfluous abundance, caught by faulconers, fowlers, fishers, and brought by my lords (Mar) tenants and purveyers to victual our campe, which consisted of fourteen or fifteen hundred men and horses.

"The manner of the hunting is this:—five or six hundred men doe rise early in the morning, and they doe disperse themselves divers wayes, and seven, eight, or ten, miles compass they doe bring or chase in the deer in many heards (two, three, or four hundred in a heard) to such or such a place as the nobleman shall appoint them; then when the day is come, the lords and gentlemen of their companies doe ride or go to the said places, sometimes wading up to the middles through bournes and rivers; and then they, being come to the place, doe lye down on the ground till those foresaid scouts, which are called the tinckell, doe bring down the deer; but as the proverb says of a bad cooke, so these tinckell men doe lick their own fingers; for besides their bows and arrows, which they carry with them, wee can heare now and then a harquebuse or musket goe off, which they doe seldom discharge in vaine: then after we had stayed three houres, or thereabouts, we might perceive the deere appeare on the hills round about us (their heads making a shew like a wood), which being followed close by the tinckell, are chased down into the valley where wee lay; then all the valley on each side being waylaid with a hundred couple of strong Irish greyhounds, they are let loose as occasion serves upon the heard of deere, that with dogs, gunnes, arrows, durks, and daggers, in the space of two houres, fourscore fat deere were slaine, which after are disposed of some one way, and some another, twenty or thirty miles; and more than enough left for us to make merrey withall at our rendevouse. Being come to our lodgings, there was such baking, boyling, rosting, and stewing as if cook Ruffian had been there to have scalded the devill in his feathers."

THE FOREST OF CORRICHIBAH.

THE forest of Corrichibah, or the Black Mount, is situated in the district of Glenorchy, in Argyllshire.

It appears from the "Black Book" (an old manuscript at Taymouth), and from other documents, to have been kept as a deer forest from a very early period, till about the time when, by the introduction of sheep on the Highland hills, the value of mountain pasture became considerably increased. At that period it ceased to be used as a forest, and was turned into sheep-farms, in which state it continued till the year 1820, when it was again converted into a forest by the present Marquis of Breadalbane.

The number of deer was at that time very small indeed, and these were scattered over a very wide district of country, namely, from the western extremity of Loch Rannoch to the head of Loch Etive on one side, and from Glencoe to Ben Aulder and Loch Eroch on the other. At this time it is not supposed that the stock of deer could have exceeded one hundred head. No sooner, however, was a part of Corrichibah kept clear from sheep, than these deer gathered in; and the number now in Lord Breadalbane's forest cannot be computed at less than 1,500. The extent of ground strictly kept for deer is about 35,000 acres. It extends on the north side from the western extremity of Loch Lydoch, by the king's house in Rannoch, to Dulness in Glen Etive; and on the south side from the confines of the county of Perth, by Loch Tulla and the River Urchay, to Corri Vicar and Glenketland. The ground is peculiarly adapted for deer, being rocky and steep, and the hills are varied with numerous corries. The rocks are mostly granite and porphyry. The grass is remarkably fine, and the sheep of the Black Mount are greatly esteemed in the Glasgow market.

The highest hills in the forest are Ben Toag, which rises on the north side of Loch Tulla; Stoupgyers, or the Hill of Goats; Clachlig, or the Stony Face; Sroin-na-forseran, or the Forester's Nose; Mealvourie, and the Craish, which rises on the south side of Glen Etive. There is a considerable extent of low ground, about nine miles in length by five or six in breadth, extending from the bases of the hills on the east side as far as Loch Lydoch. In this low ground there is a continued chain of small lochs, called the Bah Lochs, in which there are several small wooded islands; into these the deer are very fond of going. This low ground is of very great service to the forest, both as it affords good wintering and very early grass in the spring; for at that period of the year the deer may be seen standing in the water picking rushes and grass which grow at the sides of the river and lochs. This early grass is of immense importance to them, and, combined with the strong hill pasture, is one of the causes of the excellent condition in which the deer of this forest are usually found.

The stags of the Black Mount exceed those of most of the neighbouring forests in point of weight, and may be estimated at an *average* of from sixteen to seventeen stones, imperial, sinking the offal; and they are frequently found to weigh eighteen, nineteen, to twenty-one stones, having two or three inches of fat on the haunches. Their heads likewise are large in proportion, being of a much more vigorous growth than those of the Atholl or the Mar deer. One of the great advantages of the Black Mount forest is, that it forms the summit level of that part of the Highlands, and that it has equally extensive grounds on each side, both east and west; so that from whatever quarter the wind may blow, or from whatever side the deer may be disturbed, they seldom leave its bounds, but feed over either to the one side or the other. The hills being extremely rocky and precipitous, and there being only certain places by which the deer can pass from one corrie to another, the mode of killing by driving them is pretty certain. Stalking is very difficult in most parts of the forest, owing to the very steep and rugged nature of the ground. It may be mentioned, as a proof of this, that some poachers who were pursuing deer in the forest in the winter some years ago lost one of their companions, who was killed by falling over a rock.

This forest, like many others, has immemorially been believed to possess its white hind; to which, among other evidence, the following extract refers, from the old family manuscript at Taymouth, called the Black Book :—

"Upon the thettene day of February, anno 1622, the king's majesty send John Skandebar, Englishman, with other twa Englishmen in his company, to see ane quhyt hynd that was in Corrichiba, upon the 22d day of February, anno 1622."

In reference to this old story it may be mentioned that at this day there is a very light coloured deer in this forest, which all the foresters speak of as the white deer.

If "Lord Reay's country" can boast of having given birth to the celebrated poet Rob Doun, the precincts of the

"Black Mount" are not perhaps less famous for producing a bard who flourished in those rude regions about fifty or sixty years ago. His name is Duncan Macintyre; some translations from his poems have obligingly been obtained and transmitted to me by the present Marquis of Breadalbane.

Thus I have it in my power to give a specimen of the beautiful imagery of one of these translations from the Gaelic, rendered in a more modern garb by the celebrated pen of Mr. D'Israeli, jun.

SPRING IN BENDOURAN.*

Thy groves and glens, BENDOURAN, ring
With the chorus of the spring:
The blackcock chuckles in thy woods—
The trout are glancing in thy floods—
The bees about thy braes so fair,
Are humming in the sunny air;
Each sight most glad, each sound most sweet,
Amid thy sylvan pastures meet;
With the bloom of balmy May,
Thy grassy wilderness is gay!

And lo, along the forest glade
From out yon ancient pine wood's shade,—
Proud in their ruddy robes of state,
  The new-born boon of spring,
With antlered head and eye elate,
  And feet that scarcely fling
A shadow on the downy grass,
That breathes its fragrance as they pass,—
  Troop forth the regal deer:
Each stately hart, each slender hind,
Stares and snuffs the desert wind;
While by their side confiding roves
The spring-born offspring of their loves—
The delicate and playful fawn,
Dappled like the rosy dawn,
  And sportive in its fear!

The mountain is thy mother,
  Thou wild secluded race:—

---

* The inhabitants of the west still suppose that this mountain possesses the faculty of making known by strange sounds the approach of a storm, when, as they express it, "The spirit of the mountain shrieks."

Thou hast no sire, or brother,
  That watches with a face
Of half such fondness o'er thy life
Of blended solitude and strife,
As yon high majestic form
  That feeds thee on its grassy breast,
Or guards thee from the bursting storm
  By the rude shelter of its crest ;—
Or—when thy startled senses feel
  The presence of the unseen foe,
And dreams of anguish wildly steal
  O'er trembling stag, and quivering doe—
Conceals thee in her forests gloom,
And saves from an untimely doom.

Now roaming free :—for on the wind
  No sound of danger flies ;
The fawn may frolic with the hind,
  Nor fear a fell surprise ;
Or—where some knoll its verdant head
  To clustering sunbeams shows,
In graceful groups the herd may spread,
  And circling round, repose.
Thus the deer their vigils keep
Basking on Bendouran's steep !

A Poetical Translation of a Part of
"CUMHA CHORIE CHEATHARCH ;" or, The Lament for the Dell of Mist.

By a Highland Gentleman.

A TRODDEN waste each mountain side,
Whence flowed the fountain's crystal tide :
No more the grassy meads are seen,
The lovely spots of living green :
No primrose blows the silken foil ;
No herb—no floweret decks the soil
Where lay and rose the lovely hind ;
Where oft she skipped and snuffed the wind.
That hill seems now, its glory fled,
Bare as the stance of busy trade ;
Nor is the antlered monarch found—
No more he leaps with lively bound—
No more the hunter climbs the hill
To urge the forest chase with skill ;

But if there come a brighter day
To spoil the stranger of his prey,
The dell shall shine in native pride,
Shall bloom and spread its glories wide;
The stag's majestic form shall rise
Where towers yon mountain to the skies;
The roe-deer rest in shelt'ring wood—
The trout dart lively through the flood—
The hind the gentle fawn shall rear—
The hills in loveliness appear;
Each long-lost beauty bloom again
When moves the stranger from the glen.

THE FOREST OF GLENARTNEY.

The forest of Glenartney, the property of Lord Willoughby d'Eresby, is situated in a mountainous district of the same name in the parish of Comrie, and county of Perth, and contains about two thousand eight hundred Scots acres. In olden time, and even as late as 1746, it was of very considerable extent; but since that period it has been greatly reduced, and, indeed, in some measure relinquished as to forest purposes. It is bounded on the north by the Glengoinan burn, which flows eastwards into the glen that derives its name from it; and afterwards taking a northerly course, empties itself into the river Earn. On the east it is bounded by the Aultglass burn, which has its source in the mountains above Glengoinan, and is tributary to the Ruchill river. The Srathglen burn bounds it on the west, takes a southerly direction, and empties itself into the Ruchill, which forms the south boundary of the forest. The Ruchill itself rises near the high mountain Benvoirlich, about three miles west from the forest of Glenartney, and flows towards the south under the name of the Duchoran burn, until it receives many tributaries from the west and other mountain streams from the south, which rise in the hills above Doune, Dumblane, &c. Thus supplied, it becomes a formidable river, and takes the name of Ruchill (as I understand) from its rough and rocky channel. In dry weather its waters are inconsiderable; but in the stormy season it rushes with great turbulence into the Earn, and has been known to bring **down** sheep and exhausted deer along with its wreck.

There are no lakes in this forest. The chief hills **are** as follow :—Sroin-na-Cabar, Coir-na-Maville, **Ban-dhu-Boan-na-Scarnaich,** Sroin-na-Broileag, Stuic-na-Cabuic, **Beinn-Dearg,** and Sroin-na-Hellurie. **There is a** sanctuary, **or** deer-preserve, in **the centre** of **the forest,** which declines on the south, but is steep on the west, **north,** and east.

The grounds are stocked with about one hundred **black** cattle in the winter, and **one hundred and** fifty **during the** summer. The sheep were removed about seven years ago, as they were found to feed upon the best deer **pasture, and** that the **shepherds** disturbed the stags with **their** dogs. There **are perhaps from seven hundred to** one thousand **deer in the forest.** About fifty yeld hinds and forty **stags are** killed **annually, which appears to me to be a liberal proportion. As the deer are fed in the winter with corn and hay,** they attain to a considerable size. What are **called** good deer weigh, when gralloched, **from** thirteen **to fifteen** stone, and some reach even to seventeen and eighteen stone. In this forest they use both greyhounds and colley dogs for bringing wounded deer to bay ; but they seem to prefer the latter.

"The nature of the ground (says Donald Cameron, **the** old forester) is **good** and healthy, interspersed **with** heath and *rashes,* and natural grass, **and is beautiful to the** eye of the traveller." Donald has been **in the forest for** thirty-five years, and has **had the chief management of** it **nearly the** whole of that **period.**

THE FOREST OF JURA.

So common were red deer throughout Scotland, that there **are few, even of** the Hebrides, **in** which their remains are **not to be found ;** and in many of these islands, to this day, **they** still exist **in** considerable numbers. Of the latter are **Jura,** Mull, Skye, and the long **island** which includes Lewis **Harris,** North and South Uist, and Benbecula.

Whether Buchanan's derivation of the name Jura, from the Gothic word Deira, a stag, be correct, we do not pretend to say, but certain it is that in none of the Hebrides (in proportion to the extent) are deer to be found in such numbers. This island is about thirty miles in length, and five in breadth, and, with the exception of a few patches of arable land on the east coast, consists of one mountainous range extending throughout its whole length. By much the most lofty of this range are the Paps of Jura, which are situated towards the southern end of it. They are four in number, and rise from the sea on the western side, which is rugged and precipitous, and the resort of eagles and birds of prey of all sorts. The form of these hills is perfectly conical, and their elevation so abrupt, that for a considerable way from their tops no verdure is to be seen; in fact, they consist chiefly of masses of loose stone. Their height is about 2,500 feet above the level of the sea, which washes their base. The view from the top of these hills is very extensive, for, when the atmosphere is clear, the Isle of Man, and the Isle of Skye are both visible. This island is surrounded by strong tides; on the south is the rapid stream of the Sound of Islay; and on the north the famous whirlpool of Corrivrechan. The island belongs to two proprietors, Mr. Colin Campbell of Jura, and Captain MacNeill the younger, of Colonsay, whose brother has favoured me with a relation of the mode of deer-coursing practised in Jura, and already recounted in these pages. The stock consists almost entirely of sheep. The number of deer are estimated at about five hundred. They have the whole range of the island, and thus wander from one end of it to the other. As there are but few inhabitants (scarcely a thousand souls), they are seldom disturbed, and have of late years greatly increased.

From the contiguity of the sea, snow seldom lies for any length of time on these islands; and as the deer often frequent the shore, and are excessively fond of the sea ware, on which they feed even in summer, they are never altogether deprived of food, and are, consequently, much better able to endure the rigours of winter than those in a more inland situation.

The pasture in many of the valleys which intersect the island is very rich; and though there is but little brushwood, yet, from the excellence of the soil, great beds of fern are to be met with, growing to the height of six feet, in which the deer take refuge from the flies and the heat of the sun.

The district of Tarbert, beginning at the north of the loch of that name, as far as the gulf of Corrivrechan, is the part of the island most suitable to deer; the Paps are the next in estimation.

If Tarbert were cleared of sheep, and a few forest deer turned out for a cross, it would probably prove one of the finest forests in Scotland, since the pasture is excellent, the ground favourable, and the winters are mild.

When the great-grandfather of the present chief of Islay sold the island of Jura, he reserved certain forest rights, as well as others relating to the fisheries, and stipulated for a payment of six fat harts annually, and also for ten thousand oysters, as feu-duty for the holding. The chief of Islay has also a right of shooting over the island of Jura, and of taking with him such assistance as he may require. Deer, however, have been known to save him this short voyage, and to cross of their own accord to Islay, a distance of about a mile; and, in particular, six hinds and one hart did so a few years ago, and returned again to Jura. This was probably in the rutting season, and thus the hart seems to have taken a pretty effectual mode of securing to himself peaceable possession of his little seraglio.

The stags in this forest grow to a large size, and have been repeatedly killed of eighteen stones weight without the intestines. The present chief of Islay killed a hart of seventeen stones and a half Tron* weight, and in full season, whose horns were only sixteen inches from the points to the crown of the head.

---

* *Tron* weight is nearly the same as Dutch, viz., seventeen ounces and a half to the pound, and sixteen pounds to the stone; accurately speaking, perhaps, it may be a trifle more, but it is little in use.

THE ISLE OF SKYE, AND NORTH UISH.

There are about 230 deer in the Isle of Skye, which are the property of Lord Macdonald; they range over his forest near to Sconsar, and wander occasionally into the grounds of Macleod of Macleod, the other proprietor of the island. This herd has been represented to me as being in very bad plight, the full-grown stags not exceeding ten or twelve in number.

Lord Macdonald has also deer in North Uish which cannot well be got at, or followed without the assistance of boats, the island being almost entirely flat, and intersected by arms of the sea in all directions, so that there are not two miles of continuous land, and the deer, when pursued, immediately take to the water. Their number here is about 100.

LOCH ETIVE AND DALNESS.

Mr. Campbell of Monzie, whose property is situated at the head of Loch Etive, is forming a forest there, and has joined to his own lands (by lease) the old forest of Dalness, of which he is the hereditary keeper, but from which the deer have, of late years, been almost entirely expelled. By this arrangement his forest will march with Lord Breadalbane's for an extent of about six miles. Mention has already been made of a white hind referred to in the old family manuscript at Taymouth, called the Black Book, which existed in and near the forest of Corrichibah in the year 1622, and previously. It is not to be wondered at, that some superstition should attach to an animal varying so much from the natural colour of its species. Thus a tradition has been handed down in the district of Loch Etive, that, should a white hind again appear, death by violence would ensue. A few years ago (I have not received the precise date), another white hind did make her appearance, and created a great sensation on account of the above tradition. In the depth of the winter in that year, some determined poachers faced the frost and snow, when the keepers might well be supposed to be absent from the hills, and made their dispositions for driving and killing the deer. Having ascended the rugged steeps, and taken possession of the favourite passes, they sent forth their scouts to put the herd to them: these men communicated with the others, as is usual, by means of signals. As the day drew to a close, and the fading light gave a dubious appearance to the form of objects, one of the drivers who was proceeding from behind an eminence, brought his head above the sky line, and held up his arms as a signal that the deer were below. His companion in the pass, mistaking this figure for the head and horns of a stag, fired, and shot the unfortunate poacher in the head. As the whole party were engaged in an unlawful act, they endeavoured to conceal the miserable manner in which the poor fellow came by his death; so they threw the body over the rocks, which were of a great height, by which means it was so mangled, that their account of the accident, by a fall from an eminence, was very generally believed. The sister of the sufferer, however, in laying out the body, discovered the shot wound in the head, and hinted that all was not right. But as all the party had been engaged in poaching, and as the fatal occurrence was at all events an accident in which retributive justice was in no way concerned, the affair was hushed up, and is known, even at this day, but to a few.

I now conclude the catalogue and description of the forests and principal deer-haunts in the north. There may be others with which I am unacquainted; my omission to mention such (if, indeed, such do exist) will not, I trust, be imputed to my sense of their implied want of consequence, but rather to the real cause, namely, that of "pure ignorance on my part."

APPENDIX.

It may be as well to mention, that I consider the authority of the Richmond Park keeper, quoted in page 43, good only as far as it goes, and not as determining the longevity of deer in a wilder state, and under more natural circumstances.

Mr. Herring, of the New Road, London, dealer in animals, communicated a fact to me that is somewhat at variance with the authorities of Buffon and Mr. John Crerer, as mentioned in the twenty-ninth page of this volume. A full-grown hart was cut by him, which dropped his horns afterwards, and had fresh ones the succeeding season of 1838. This hart I myself saw, but the new horns were misshapen and diminutive.

THE HIGHEST HILLS IN THE FOREST OF ATHOLL.

Felaar Forest.

Glastullich.
Benuirn.
Cairnanree.
Malnaspionach.
Cairn Dairg.
Gailcharn.

Hell's Hill.
The King's Cairn.

The Red Cairn.
The White Cairn.

Ben-y-Gloe Forest.

Cairn-na-Gour.
Argiotvane.
Cairnicklechalm.
Ben V.eg.
Ben Vourich.
Cairn Lia.
Cairn Torkie.

Goat's Hill.

Little Hill.
Boar's Hill.
Grey Cairn.
The Boar's Cairn.

Malvourich.
Garorune.
Top of Carrie Chastail.   The Castle Hill.

South Side of Tarff.

Sligernoch.
Monlia.   The Grey Hill.
Conaloch.
Maltenail.   The Gathering Knoll.
Benchroam.   The Crooked Hill.
Craig croachie.   The Hanging Rock.
Grennach.
Cairnchlamain.   The Glead's Cairn.
Sroin a Chro.
Cairn-Maronach, or Cuirn-Marnich. Braemar Cairns.
Glas Mal.   Grey Knoll.
Ben-y-Venie.   The Middle Hill.
Ben Chat.   Hill of the Cat.
Mallour.   Dun-coloured Knoll.
Cairn Cherrie.
Elerick.
Ben Derig.   The Red Hill.
Torr.
Ben Toaskernich.   The Toad's Hill.
Craig na Helleir.   The Eagle's Rock.

North Side of Tarff.

Corrie na Craig.
Malcrapan Laagh.   The Knoll of the Calves.
Malna Cairn.   The Knoll of the Cairn.
Mackaranoch.
Scarsach.
Malduchlach.   The Knoll of the Blackstone.
Malcuirn.   Knoll of the Cairn.
Corrie Stock Guise.
Cairnan Illair.   Fiddlers Cairn.
Drimliafeaheaskichan.
Mal Glashea.
Mal **Corrie** Vreak.
Mal Corriechraggach.
Mal dubh na Glashea.
Ben Vreak.
Sligernoch.
Corrie crom na damk.
Druim **Corrie na Rislechan.**
Druim na feachanouer.
Druim Minagag.
Uchg na Clasair.

Glen Garry Forest.

Mal cham corrie.
Sroin feachon.
Drium Kirrichon.
Mal ouer.
Sroin a chlerick.
Mal voulin.
Mal Vrammie.
Sroin Glasechorrie.
Vi chosach.
Sroin Craig an Loch.
Dune.
Cricharickrior.
Medher.
Glaish Mal east.
Glaish Mal west.
Monadh Lia.
Vuinnach.
Corrie Mac Shee.
Craig Chursech.

South Side of Inverness Road.

Mackrannoch.
Tork, or The Duke of Atholl's Boar.
Mal Corrie Vackie.
Mal dourune.
Ben Derig, top of Corrie Lunnie.
Carkel Lock garry.
Mal na Letirch.
Mona baan.
Carkel.*

EVIDENCE *relating to the* TRIAL *of* DUNCAN TERIG, *alias* CLERK, *and* ALEXANDER BAIN MACDONALD, *for the* MURDER *of* SERGEANT DAVIES.

ALEXANDER M'PHERSON alias M'Gillas, in Inverey, being solemnly sworn, purged of malice and partial council and interrogate, aged twenty-six years, unmarried, deposes, that in summer, 1750, he found lying in a moss bank on the hill of Christie, a human body,

---

* I have given these names in the most correct local orthography I could obtain, but no two people spell all of them precisely in the same manner; many of them, indeed, are so corrupted, that their very meaning is lost: this, perhaps, may have in a great measure originated from the uncomplying pronunciation of strangers. Thus they write Ben-derig, Ben Derg, and Ben dairg. In this dilemma I have thought it best to make use of the name most generally received.

at least the bones of a human body, of which the flesh was mostly
consumed, and he believed it to be the body of Sergeant Davies,
because it was reported in the country that he had been murdered
on that hill the year before; that when he first found this body,
there was a bit of blue cloth upon it pretty entire, which he took
to be what is called English cloth; he also found the hair of the
deceased, which was of a dark mouse colour, and tied about with
a black ribbon; that he also observed some pieces of a striped
stuff; and also found lying there a pair of brogues, which had
been made with latches for buckles, which had been cut away
by a knife; that he, by help of his staff, brought out the body,
and laid it upon plain ground, in doing whereof, some of the bones
were separated one from another; deposes, that for some days he
was in doubt what to do, but meeting with John Grawar in the
moss, he told John what he had found, and John bid him tell
nothing of it, otherwise he would complain of the deponent to
John Shaw of Daldownie, upon which the deponent resolved to
prevent Grawar's complaint, and go and tell Daldownie of it
himself; and which having accordingly done, Daldownie desired
him to conceal the matter, and go and bury the body privately,
as it would not be carried to a kirk unkent, and that the same
might hurt the country, being under the suspicion of being a rebel
country; deposes, that some few days thereafter, he acquainted
Donald Farquharson, the preceding witness, of his having seen
the body of a dead man in the hill, which he took to be the body
of Sergeant Davies; that Farquharson at first doubted the truth
of his information, till the deponent told him that a few nights
before, when he was in bed, a vision appeared to him, as of a man
clad in blue, who told the deponent, "I am Sergeant Davies;"
but that before he told him so, the deponent had taken the said
vision at first appearance to be a real living man, a brother of
Donald Farquharson; that the deponent rose from his bed, and
followed him to the door, and then it was, as has been told, that
he said he was Sergeant Davies, who had been murdered in the
Hill of Christie, nearly about a year before, and desired the
deponent to go to the place he pointed at, where he would find
his bones; and that he might go to Donald Farquharson, and
take his assistance to the burying of him; that upon giving
Donald Farquharson this information, Donald went along with
him, and finding the bones as he had informed Donald, and
having then buried them with the help of a spade, which he, the
deponent, had along with him; and for putting what is above
deposed upon out of doubt, deposes, that the above vision was the
occasion of his going by himself to see the dead body; and which
he did before he either spoke to John Grawar, Daldownie, or any
other body; and further deposes, that while he was in bed,
another night, after he had first seen the body by himself, but
had not buried it, the vision again appeared naked, and minded
him to bury the body; and after that he spoke to the other folks

above mentioned, and at last complied, and buried the bones above mentioned: deposes, that upon the vision's first appearance to the deponent in his bed, and after going out of the door, and being told by it (the vision) that he was Sergeant Davies, the deponent asked him who it was that had murdered him, to which it made this answer, that if the deponent had not asked him, he might have told him, but as he had asked him, he said he either could not, or would not, but which of the two expressions the deponent cannot say; but at the second time the vision made its appearance to him, the deponent renewed the same question, and then the vision answered, that it was the two men now in the panel who had murdered him; and being further interrogated in what manner the vision disappeared from him first and last, deposes, that after the short interviews above-mentioned, the vision at both times disappeared and vanished out of his sight in the twinkling of an eye; and that in describing the panels by the vision above mentioned, as his murderers, his words were Duncan Clerk and Alexander Macdonald: deposes, that the conversation betwixt the deponent and the vision was in the Irish language: deposes, that several times in the harvest before the Martinmas, after seeing the said vision, he was applied to by Duncan Clerk, the panel, then to enter home to his service at that time, which accordingly he did, and staid in his service just a year; and he being on the hill together with Duncan Clerk, spying a young cow, desired the deponent to shoot it; and though Duncan did not bid him carry it home after it should be shot; yet the deponent understood that to be the purpose, when Duncan desired him to shoot it, and which the deponent refused to do, adding, that it was such thoughts as these were in his head when he murdered Sergeant Davies; upon which some angry expressions happened between Duncan and the deponent; but when the deponent insisted upon it that he could not deny the murder, Duncan fell calm, and desired the deponent to say nothing of that matter, and that he would be a brother to him, and give him every thing he stood in need of, and particularly would help him to stock a farm when he took one. At the time of deposing, the deponent exhibited a paper, which is marked on the back by the Lord Examiner, the deponent averring that he cannot write, and deposes, that the said paper was put into his hands by the said Duncan Clerk, who at the time told him it was a premium of twenty pounds Scots to hold his tongue of what he knew of Sergeant Davies: deposes, that while the deponent was in the panel Clerk's service, and about Lammas, 1751, he showed to the deponent a long green silk purse, and that he showed, also, to the deponent the contents which were in it, namely, sixteen guineas in gold, and some silver; and being interrogate what was the occasion of showing this purse and money to the deponent, deposes—it was one of two which he does not remember—either he had come from Aberdeen with money, which he had got for his wool, or was going to Badenoch to buy sheep; deposes, that he saw upon the finger of Elizabeth Downie, the panel Duncan Clerk's wife, a yellow ring, which she told him was gold, with a plate on the outside of it in the form of a seal, and that he saw it on her finger six or eight weeks before her marriage; and that after her marriage, she having one day taken it off her finger, he saw upon the inside of it a stamp, but what that stamp is he does not know; and being interrogate, deposes, that he had a suspicion that this ring was Sergeant Davies's ring, having heard it reported in the country that Sergeant Davies had such a ring upon his finger when he was murdered, but does not remember his having told his suspicion to anybody; and being further interrogate, deposes, that since the panel Duncan's imprisonment, the deponent was solicited by Donald Clerk, the panel Duncan's brother, to conceal what he knew when he came to give evidence; but this was after his having first solicited the deponent to leave the country, that he might not give evidence, and upon the deponent's saying he offered him nothing to leave the country with; but then it was that Donald proposed his not giving true evidence, adding, that of every penny Donald was worth, the deponent should have the half; and being interrogated, at the desire of the jury, if ever he had asked payment of the twenty pounds contained in the above-mentioned paper produced by him, deposes, that he once did, shortly after the term of payment, to which Duncan answered that it would be as well to let it lie in his hands, with which he was satisfied; and that he never asked payment of the annual rent; and being further interrogate, deposes, that before the deponent went home to the panel's service, at Martinmas, 1750, it was well known and reported in the country that the bones of the dead body found upon the above-mentioned hill had been buried by the deponent and Donald Farquharson, as also was the story of the vision or apparition, whereof the deponent had told Donald Farquharson; and being interrogate for the panel, deposes, that he not only told the story of the vision, or apparition to Donald Farquharson, as above mentioned, but that he also told it to John Grawar and Daldownie before he mentioned it to Donald Farquharson; deposes that there were folks living with him at the sheiling, when the vision appeared to him as above, but that he told it to none of them; and adds that Isabel M'Hardie, in Inverey, a woman then in the sheiling with him, has told him since, that she saw such a vision as the deponent has above described, and has told him herself so much; and upon the panel's interrogatory, deposes, that upon the vision's appearing to him, it described the place where he would find the bones so exactly, that he went within a yard of the place where they lay upon his first going out: and this is the truth, as he shall answer to God; and deposes he cannot write.

(Signed)   JA. FERGUSON.

Isabel M'Hardie of Inverey also gave solemn evidence of her having seen the apparition, having deposed "that one night, about four years ago, when she was lying at one end of the sheiling, and Alexander Macpherson, who was then her servant, lying in the other, she saw something naked come in at the door, which frighted her so much, that she drew the clothes over her head; that when it appeared, it came in a bowing posture; and that next morning she asked Macpherson what it was that had troubled them the night before, to which he answered, she might be easy, for that it would not trouble them any more."

THE END.

www.ingramcontent.com/pod-product-compliance
Lightning Source LLC
Chambersburg PA
CBHW030317240426
43673CB00040B/1198